30-SECOND
MATH

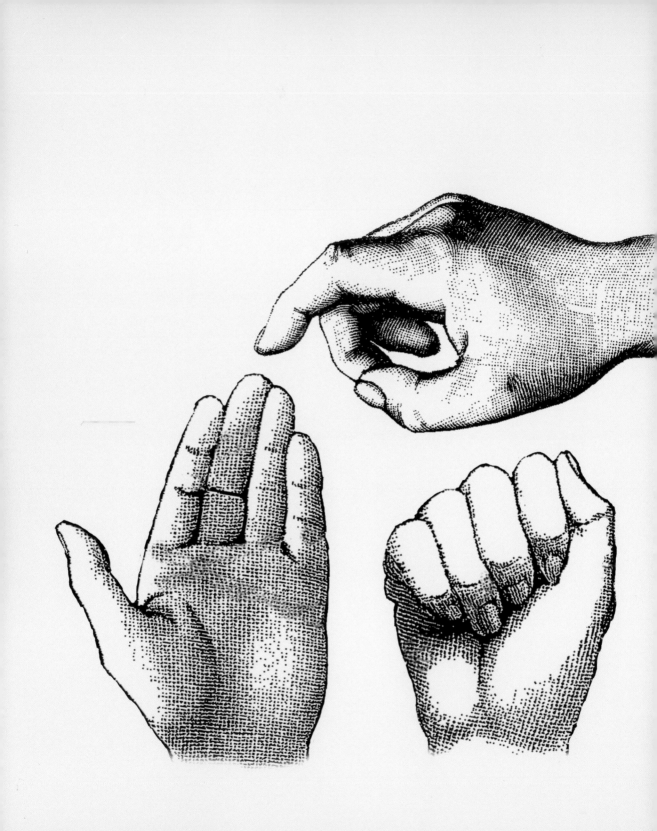

30-SECOND
MATH

The 50 most mind-expanding
theories in mathematics, each
explained in half a minute

Editor
Richard Brown

Contributors
Richard Brown
Richard Elwes
Robert Fathauer
John Haigh
David Perry
Jamie Pommersheim

METRO BOOKS
New York

METRO BOOK
New York

An Imprint of Sterling Publishing
1166 Avenue of the Americas
New York, NY 10036

This book was conceived,
designed, and produced by

Ivy Press
210 High Street, Lewes,
East Sussex BN7 2NS, U.K.
www.ivypress.co.uk

Creative Director **Peter Bridgewater**
Publisher **Jason Hook**
Editorial Director **Caroline Earle**
Art Director **Michael Whitehead**
Designer **Ginny Zeal**
Illustrator **Ivan Hissey**
Profiles Text **Viv Croot**
Glossaries Text **Steve Luck**
Project Editor **Jamie Pumfrey**

ISBN: 978-1-4351-5261-8

For information about custom
editions, special sales, and premium
and corporate purchases, please contact
Sterling Special Sales at 800-805-5489
or specialsales@sterlingpublishing.com.

Manufactured in China

Color origination by Ivy Press Reprographics

2 4 6 8 10 9 7 5 3

www.sterlingpublishing.com

CONTENTS

INTRODUCTION
Richard Brown

It is said that mathematics is the art of pure
reason. It is the fundamental logical structure of all that exists,
and all that doesn't exist, in this reality of ours. Far removed from
the simple calculations that allow us to balance our accounts and
calculate our everyday affairs, mathematics helps us to order and
understand the very notion of everything we can imagine in life.
Like music, art and language, the essential symbols and concepts
of mathematics, many of which are defined and discussed in this
book, allow us to express ourselves in amazingly intricate ways and
to define unimaginably complex and beautiful structures. While the
practical uses for mathematics are rife, what makes mathematics
so magical is its elegance and beauty outside of any real application.
We give the concepts in mathematics meaning only because they
make sense and help us to order our existence. But outside of the
meaning we give these elements of math, they do not really exist at
all except in our imagination.

The natural and social sciences use mathematics to describe their
theories and provide structure for their models, and arithmetic and
algebra allow us to conduct our business and learn how to think.
But beyond these practical applications lies the true nature of the
discipline; mathematics is the framework of and provides the rules
of engagement for the entire system of structured thought.

This text is a glimpse into the world a mathematician sees in
everyday life. Herein lies a set of some of the more basic and
fundamental elements in the field today, with definitions, a
little history, and some insight into the nature of many basic

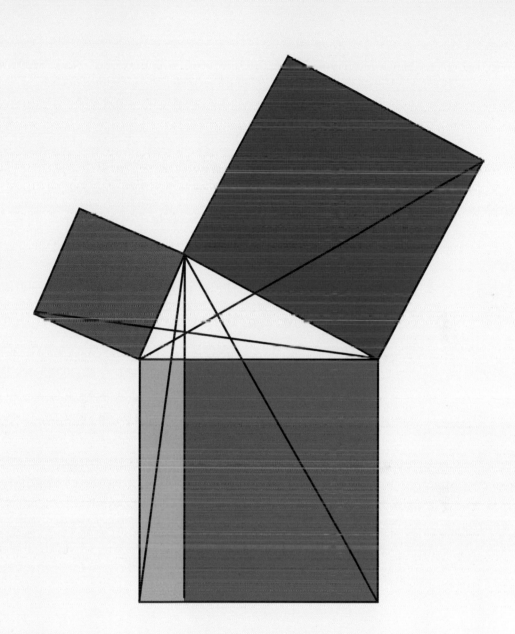

Elegant geometry
*Mathematicians often "see"
mathematical objects like
equations using geometry.
This is a visual proof of the
famous Pythagorean
Theorem, $a^2+b^2=c^2$.*

mathematical concepts. This book contains 50 entries, each of which centers on an important topic in mathematics. They are ordered into seven categories, which roughly help to define their context. In **Numbers & Counting**, we explore the basic building blocks that allow us to enumerate our surroundings. We study some of the operations and structures on numbers in **Making Numbers Work**. These entries basically describe the arithmetic system that enables us to use mathematics in our everyday lives. In **Chance is a Fine Thing**, we detail some ideas and consequences that arise when using mathematics to understand random events and chance happenings. Next, we lay out some of the deeper, more complex structures of numbers in **Algebra & Abstraction**. It is here that the path toward higher mathematics begins. In turn, we explore the more visual aspects of mathematical relationships in **Geometry & Shapes**. Since mathematical abstraction is one of pure imagination, we then explore what happens outside of our three dimensions in **Another Dimension**. And finally, in **Proofs & Theorems**, we discuss some of the more profound ideas and facts to which our mathematical path has led us.

Individually, each entry is a brief glimpse into one of the more beautiful and important ideas central to mathematics today. Each topic is presented in the same format, aimed at facilitating a proper introduction; the 3-second sum offers the briefest overview, the 30-second math goes into further depth on the topic, and a 3-minute addition begins the process of pondering the deeper connections between the idea and its importance in the world. It is hoped that, taken together, these elements will help to open your eyes to a deeper understanding of the nuts and bolts of what mathematics is really all about.

When used as a reference text, this book will provide the basics of some of the more profound ideas in mathematics. When read in full, this text may provide a glimpse into another world as rich and meaningful as the one you live in now: the world of mathematics.

Dimensional beauty
There are only five ways to construct a three-dimensional solid using regular polygons. It is not hard to see why. But does that make these objects special? Mathematicians think so.

NUMBERS & COUNTING

algebra One of the main branches of pure mathematics that studies operations and relations on numbers. Elementary algebra involves studying the rules of arithmetic on expressions involving variables. Advanced algebra involves studying these operations and relations on mathematical objects and constructions other than numbers.

algebraic number Any number that is a root of a non-zero polynomial that has integer coefficients. In other words algebraic numbers are solutions to polynomial equations (see page 80), such as $x^2 - 2 = 0$, where $x = \sqrt{2}$. All rational numbers are algebraic, but irrational numbers can be either algebraic or not. One of the best-known algebraic numbers is the golden ratio (1.6180339...), which is usually written ϕ.

binary (base 2) The counting system in which only the numbers 1 and 0 feature. Just as in our base 10 system there is a 1s column ($10^0 = 1$), 10s column (10^1) and 100s (10^2) column, and so on, in base 2 there is a 1s (2^0) column, a 2s column ($2^1 = 1$), a 4s column (2^2), and so on. For example, the binary version of 7 is written 111, as in $1 \times 1 + 1 \times 2 + 1 \times 4$.

coefficient A number that is used to multiply a variable; in the expression $4x = 8$, 4 is the coefficient, x is the variable. Although usually numbers, coefficients can be represented by symbols such as a. Coefficients that have no variables are called constant coefficients or constant terms.

complex number Any number that comprises both real and imaginary number components, such as $a + bi$, in which a and b represent any real number and i represents $\sqrt{-1}$. See *imaginary number*.

factor One of two or more numbers that divides a third number exactly. For example 3 and 4 are factors of 12, as are 1, 2, 6, and 12.

figurate number Any number that can be represented as a regular geometric shape, such as a triangle, square, or hexagon.

fractional number (fraction) Any number that represents part of a whole. The most common fractions are called common or vulgar fractions, in which the bottom number, the denominator, is a non-zero integer denoting how many parts make up the whole, whereas the top number, the numerator, represents the number of equal divisions of the whole. Proper fractions represent a value of less than 1, for example, $^2/_3$, whereas improper fractions represent a value greater than 1, for example, $^3/_2$, or $1^1/_3$.

imaginary number A number that when squared provides a negative result. As no real number when squared provides a negative result, mathematicians developed the concept of the imaginary number unit i, so that $i \times i = -1$ or put another way $i = \sqrt{-1}$. Having an imaginary number unit that represents $\sqrt{-1}$ helps solve a number of otherwise unsolvable equations, and has practical applications in a number of fields.

integer Any natural number (the counting numbers 1, 2, 3, 4, 5, and so on), 0, or the negative natural numbers.

irrational number Any number that cannot be expressed as a ratio of the integers on a number line. The most commonly cited examples of irrational numbers are π and $\sqrt{2}$. A good way of identifying an irrational number is to check that its decimal expansion does not repeat. Most real numbers are irrational numbers.

number line The visual representation of all real numbers on a horizontal scale, with negative values running indefinitely to the left and positive to the right, divided by zero. Most number lines usually show the positive and negative integers spaced evenly apart.

polynomial An expression using numbers and variables, which only allows the operations of addition, multiplication, and positive integer exponents, i.e., x^2. (See also Polynomial Equations, page 80.)

rational number Any number that can be expressed as a ratio of the integers on a number line; or, more simply, any number that can be written as a fraction, including whole numbers. Rational numbers are also identified by finite or repeating decimals.

real number Any number that expresses a quantity along a number line or continuum. Real numbers include all of the rational and the irrational numbers.

transcendental number Any number that cannot be expressed as a root of a non-zero polynomial with integer coefficients; in other words non-algebraic numbers. π is the best known transcendental number, and following the opening definition π therefore could not satisfy the equation $\pi^2 = 10$. Most real numbers are transcendental.

whole number Also known as a natural or counting number, a whole number is any positive integer on a number line or continuum. Opinion varies, however, on whether 0 is a whole number.

FRACTIONS & DECIMALS

the 30-second math

The whole numbers, 0, 1, 2, 3 ...,
are the bedrock of mathematics, and have
been used by humans for millennia. But not
everything can be measured using whole
numbers. If 15 acres of land are divided
between 7 farmers, each will have $^{15}/_7$ (or $2^1/_7$).
The simplest non-whole numbers can be
expressed in a fractional form like this. But
for other numbers, such as π, this is awkward
or impossible. With the development of science
came the need to subdivide quantities ever
more accurately. Enter the decimal system,
an efficient column-based method using
Hindu-Arabic numerals. Here, the number 725
has three columns, and stands for 7 hundreds,
2 tens, and 5 units. By adding a decimal point
after the units, and extra columns to its right,
this approach easily extends to numbers smaller
than a unit. So, 725.43 stands for 7 hundreds,
2 tens, 5 units, 4 tenths (of a unit) and 3
hundredths. By incorporating ever more columns
to the left or to the right, numbers both large
and small can be written as precisely as needed.
In fact, every number in between the whole
numbers can be expressed as a decimal (but not
as a fraction), giving us the "real" number
system.

RELATED THEORIES
See also
RATIONAL & IRRATIONAL
NUMBERS
page 16
COUNTING BASES
page 20
ZERO
page 36

3-SECOND BIOGRAPHIES
ABU 'ABDALLAH MUHAMMAD
IBN MUSA AL-KHWARIZMI
c. 770–c. 850

ABU'L HASAN AHMAD IBN
IBRAHIM AL-UQLIDISI
c. 920–980

IBN YAHYA AL-MAGHRIBI
AL-SAMAWAL
c.1130–1180

LEONARDO PISANO
(FIBONACCI)
c. 1170–c. 1250

30-SECOND TEXT
Richard Elwes

*Whole numbers can
be subdivided into
fractions, and decimals
express these divisions
even more precisely.*

3-SECOND SUM
The starting point for
mathematics is the system
of whole numbers, 0, 1, 2,
3... But many things fall
between the gaps, and
there are two ways to
measure them.

3-MINUTE ADDITION
Translating between
fractions and decimals is
not always straightforward.
It is easy to recognize
0.25, 0.5, and 0.75 as ¼,
½, and ¾ respectively.
But the decimal equivalent
of ⅓ is 0.333333...,
where the string of 3s
never ends, and ⅐ is
0.142857142857142857...,
also with a never-ending
repeating pattern. It turns
out all fractional numbers
have repeating patterns
in their decimal, while
non-fractional numbers
like π have decimals that
do not repeat. These are
the irrational real numbers.

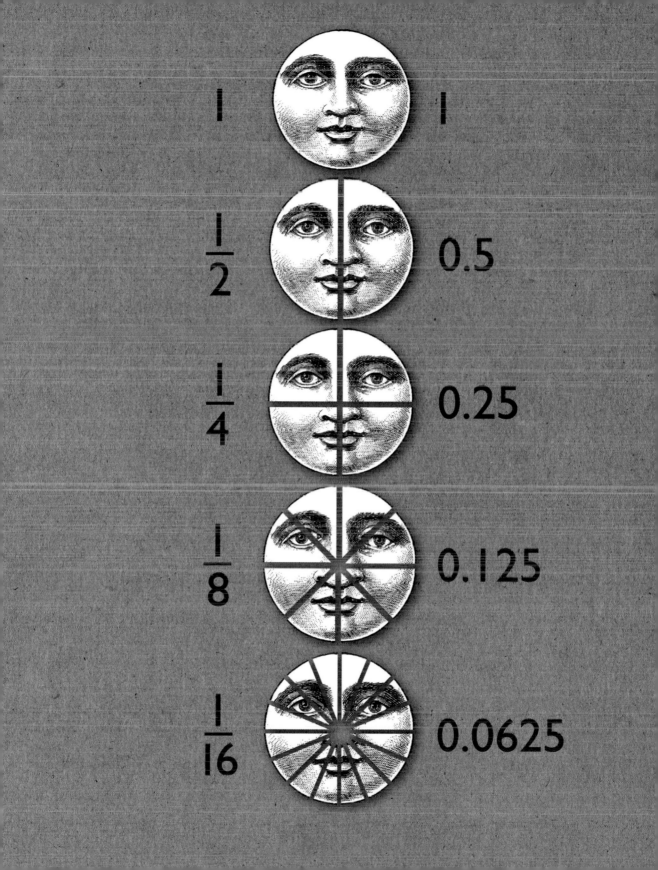

1   1

1/2   0.5

1/4   0.25

1/8   0.125

1/16   0.0625

RATIONAL & IRRATIONAL NUMBERS

the 30-second math

3-SECOND SUM
"Real" numbers—the numbers used to express quantities and representable via a decimal expansion—are either rational or irrational. But some irrationals are more unusual than others.

3-MINUTE ADDITION
The philosophy of the ancient Greeks held that all things measurable are, at worst, the ratio of whole numbers. Anecdotal history holds that the Pythagoreans were so distraught to discover that $\sqrt{2}$ is irrational that Hippasus of Metapontum was murdered to prevent revelation of this truth to the world. A number like π is perhaps more intuitively irrational, but it was only 250 years ago that this was proved true, and another century would pass before π was proved to be transcendental.

Real numbers consist of positive numbers, negative numbers and 0, and these values can be categorized in several ways. The most fundamental way is to distinguish the real numbers that can be expressed as the fraction of two integers, such as $\frac{1}{2}$ or $-\frac{7}{3}$ (called rational numbers), from those that cannot (called irrational numbers). The ancient Greeks believed all numbers were rational, until a follower of Pythagoras proved that $\sqrt{2}$ is not rational. You can tell if a number is rational or irrational by looking at its decimal expansion— if the digits ultimately repeat, the number is rational (think $\frac{3}{11} = 0.272727...$). Decimal expansions of irrational numbers (for example, $\pi = 3.14159265...$) have digits that do not repeat. But there's more. Rational numbers and many irrational numbers have something in common – they are algebraic, that is they are solutions to polynomial equations with integer coefficients. For example, $\sqrt{2}$ solves $x^2 - 2 = 0$ (see Polynomial Equations, page 80). But many more irrational numbers are not algebraic, and π is one example. Numbers that are not algebraic are called transcendental—only irrational numbers can be transcendental.

RELATED THEORIES
See also
FRACTIONS & DECIMALS
page 14
EXPONENTIALS
& LOGARITHMS
page 44
POLYNOMIAL EQUATIONS
page 80
PI—THE CIRCLE CONSTANT
page 96
PYTHAGORAS
page 100

3-SECOND BIOGRAPHIES
HIPPASUS OF METAPONTUM
active fifth century BCE

JOHANN LAMBERT
1728–1777

CHARLES HERMITE
1822–1901

FERDINAND VON LINDERMANN
1852–1939

30-SECOND TEXT
David Perry

Be real—numbers are rational if they can be written as a fraction. Otherwise they are irrational.

IMAGINARY NUMBERS

the 30-second math

3-SECOND SUM
Today's mathematicians work in an expanded number system, which includes a new "imaginary" number i, the square root of -1.

3-MINUTE ADDITION
The complex numbers allow for solutions to equations like $x \times x = -1$. One might ask next whether there are solutions to $x \times x = i$, for example, or whether we have to expand the system yet again. As it turns out, the complex numbers contain solutions to all possible polynomial equations, meaning that they are all we will ever need. This wonderful fact is known as the fundamental theorem of algebra.

Over the years, mathematicians have enlarged the number system several times. An early expansion was the inclusion of negative numbers. In business, for example, if $+4$ represents being in profit by 4 units, then -4 stands for being 4 units in debt. Negative arithmetic has a surprising property. Multiply a positive number by a negative, and you get a negative result: e.g., $-4 \times 3 = -12$. But multiply one negative number by another, and you get a positive result: $-4 \times -3 = 12$. So there was no number (positive or negative) that, when multiplied by itself, gives a negative answer. This meant that some simple equations, such as $x^2 = -1$, could never be solved, which was an obstacle to solving more sophisticated equations, even when solutions existed. This was corrected by a new "imaginary" number i, defined as the square root of -1; that is to say $i \times i = -1$. This started off as a cheat to assist in calculations and was controversial early on; Descartes coined the term "imaginary" as a derogatory term. Over time, however, it has become as accepted as all other types of number. Today, the number system that mathematicians prefer is termed "complex numbers," comprising the likes of $2 + 3i$, or $\frac{1}{2} - \frac{1}{4}i$, or more generally $a + bi$, where a and b are any "real" (that is to say decimal) numbers.

RELATED THEORIES
See also
FRACTIONS & DECIMALS
page 14
POLYNOMIAL EQUATIONS
page 80
RIEMANN'S HYPOTHESIS
page 150

3-SECOND BIOGRAPHIES
NICCOLÒ FONTANA
('TARTAGLIA')
1500–1557

GIROLAMO CARDANO
1501–1576

RAFAEL BOMBELLI
1526–1572

CARL-FRIEDRICH GAUSS
1777–1855

AUGUSTIN-LOUIS CAUCHY
1789–1857

30-SECOND TEXT
Richard Elwes

Positive and negative integers weren't enough for some mathematicians— they needed imaginary numbers.

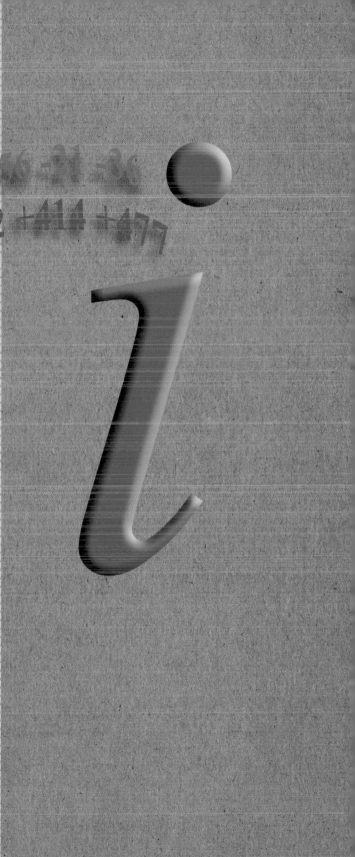

COUNTING BASES

the 30-second math

RELATED THEORY
See also
ZERO
page 36

3-SECOND BIOGRAPHIES
GOTTFRIED LEIBNIZ
1646–1716

GEORGE BOOLE
1815–1864

30-SECOND TEXT
Richard Brown

3-SECOND SUM
A base refers to the number of unique digits that a counting system uses to represent numerical values.

3-MINUTE ADDITION
The Mayans of Central America also used base 20 for the "long count" of their calendar, although they "corrected" the third column from the normal $400 = 20 \times 20$ part to $18 \times 20 = 360$, maybe to reflect the approximate number of days in a year. If we prefer base 10 simply because our fingers are good calculators, did the Mayans see the value of their open-shoed toes in this endeavor?

When we count numbers beyond nine, we are used to putting a "1" in the next column and reusing the symbols. This is because we use the base 10 or decimal system. But base 10 has not always been the preferred system. Ancient Babylonians used base 60 (the sexagesimal system) for counting. Rather than stopping at nine and moving into the next column, they stopped at 59. Some reminders of this system include the continued use of 60 minutes in an hour, and 360° in a circle. References to base 12 counting (the duodecimal system) give us the concepts of dozen and gross (a dozen dozen). Base 20 counting (the vigesimal system) was common in early Europe (the "score" in Abraham Lincoln's famous Gettysburg Address line, "4 score and 7 years ago," is 20). Modern computers use the base 2 or binary number system, where only 0 and 1 are used. Here it was easy to produce early systems for counting where only two mutually exclusive states are needed, like an open or closed electrical circuit. In any base, addition and multiplication are well defined and one can do algebra. Try that the next time someone asks you for the value of 1 plus 1. It is obviously 10 (in binary arithmetic)!

The most commonly used counting system is base 10—the Babylonians thought big with 60 unique digits. Computer code keeps it simple with a mere two digits.

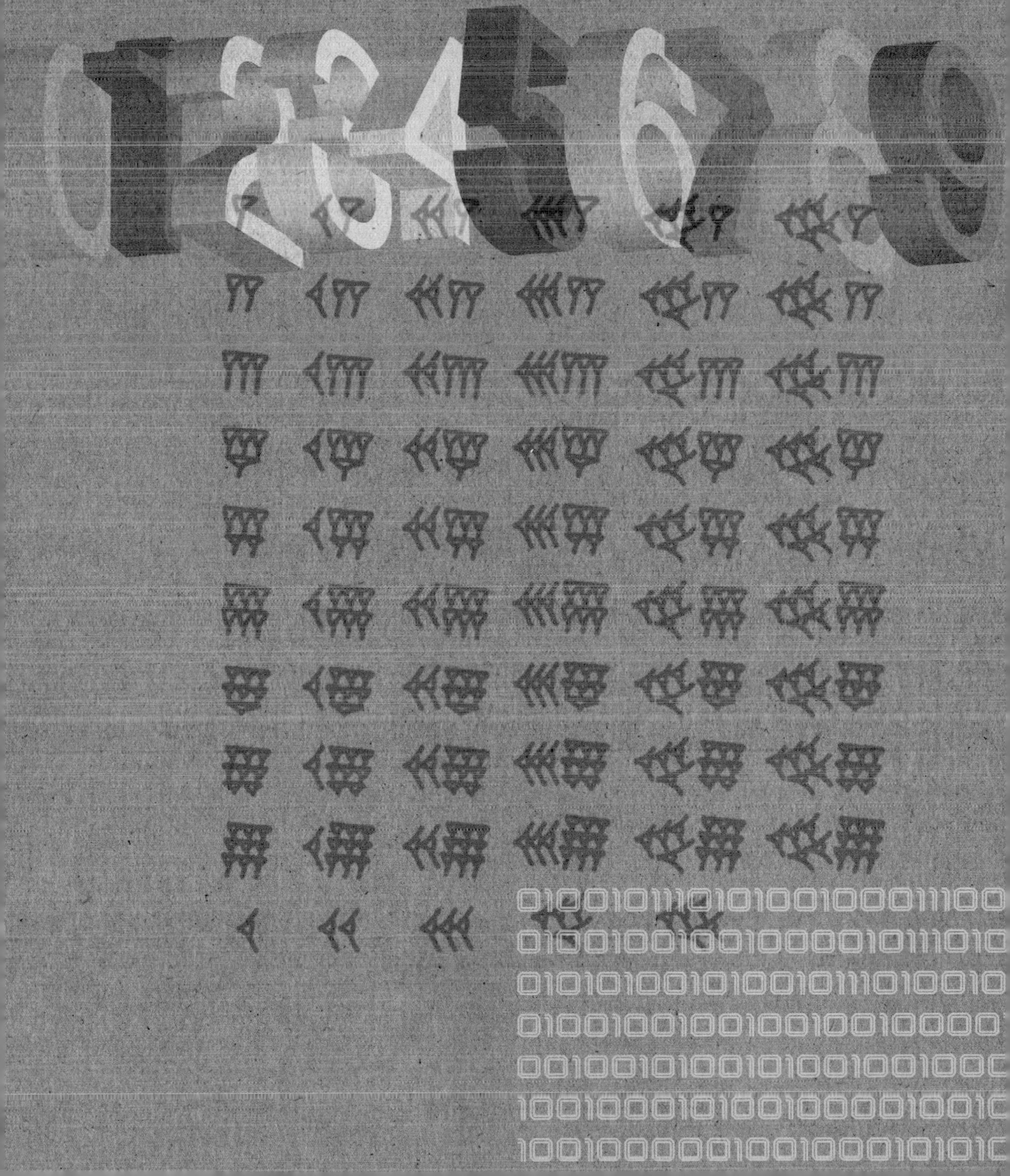

PRIME NUMBERS

the 30-second math

Most whole numbers will factor
into smaller parts. For example, $100 = 4 \times 25$.
It's also true that $100 = 20 \times 5$. If we take
either of those and break the factors into still
smaller factors, we ultimately come to the prime
factorization of $100:100 = 2 \times 2 \times 5 \times 5$.
We cannot break down the factors further—
they are prime, divisible only by 1 and
themselves. When mathematicians started
listing the prime numbers, they searched for a
pattern but did not see one. They raised the
question of whether the list was finite or if one
could find larger and larger primes. Euclid gave
an elegant proof in his *Elements* that there are
infinitely many primes. 17,463,991,229 is a
large prime. How do we know it's prime? We
could try dividing this integer by all smaller
integers and find no factors other than 1, then
declare it prime. This is slow, however, and there
are better ways. The largest known primes have
more than 10,000,000 digits, and clever
methods are required to establish them as such.
Finding large primes might seem frivolous, but a
revolutionary idea in the 1970s created a
technique to effect secure communications by
use of a system requiring the generation of large
prime numbers. This technique pervades the
Internet, allowing us to shop online in safety.

*Only divisible by 1
and themselves,
prime numbers
have fascinated
mathematicians
for centuries. The
discovery of large
primes has practical
applications today.*

1	2	3	4	5	6	7	8	9	10
11	12	13	14	15	16	17	18	19	20
21	22	23	24	25	26	27	28	29	30
31	32	33	34	35	36	37	38	39	40
41	42	43	44	45	46	47	48	49	50
51	52	53	54	55	56	57	58	59	60
61	62	63	64	65	66	67	68	69	70
71	72	73	74	75	76	77	78	79	80
81	82	83	84	85	86	87	88	89	90
91	92	93	94	95	96	97	98	99	100

FIBONACCI NUMBERS

the 30-second math

In the Fibonacci sequence

1, 1, 2, 3, 5, 8, 13, 21, 34, 55, 89, 144, 233, ... each term is the sum of the previous two terms. The resulting sequence, which plays a special role in number theory, possesses many curious numerical properties. If you add the terms in the Fibonacci sequence up to a certain point, the sum is always one less than a Fibonacci number; e.g., $1 + 1 + 2 + 3 + 5 + 8$ is one less than the Fibonacci 21. Adding the squares of these numbers produces a product of two Fibonacci numbers: $1 + 1 + 4 + 9 + 25 + 64 = 8 \times 13$. The ratios 1:1, 2:1, 3:2, 5:3, 8:5, ... approach the golden ratio $\phi \approx 1.618$. Geometrically, squares whose sides are Fibonacci numbers in length fit together nicely to form a golden spiral. Long before humans became fascinated with these patterns, plants had discovered the economy of Fibonacci numbers. The leaves or buds of many plants with a spiral structure—such as pineapples, sunflowers and artichokes—exhibit a pair of consecutive Fibonacci numbers. Examining a pineapple, you'll find 8 rows spiraling around in one direction and 13 in the other direction. In the animal kingdom, a honeybee has a Fibonacci number of ancestors in each generation.

3-SECOND SUM
A simple rule, adding the two previous terms to get the next term, produces one of Mother Nature's favourite sequences of numbers.

3-MINUTE ADDITION
In 1202, Leonardo Pisano, also known as Fibonacci, posed a riddle about breeding rabbits in his book *Liber Abaci* (*The Book of the Abacus*). Fibonacci posited, perhaps unrealistically, that after every month, each pair of adult rabbits produces one pair of baby rabbits, and baby rabbits take one month to become adults. If you start with a single pair of baby rabbits in January, you will have 144 pairs of rabbits by December!

RELATED THEORIES
See also
NUMBER THEORY
page 30
THE GOLDEN RATIO
page 98

3-SECOND BIOGRAPHY
LEONARDO PISANO
(FIBONACCI)
C. 1170–C. 1250

30-SECOND TEXT
Jamie Pommersheim

Fibonacci numbers appear in the ancestral tree of a honeybee. Each male bee has only a female parent, while each female has two parents, one male and one female.

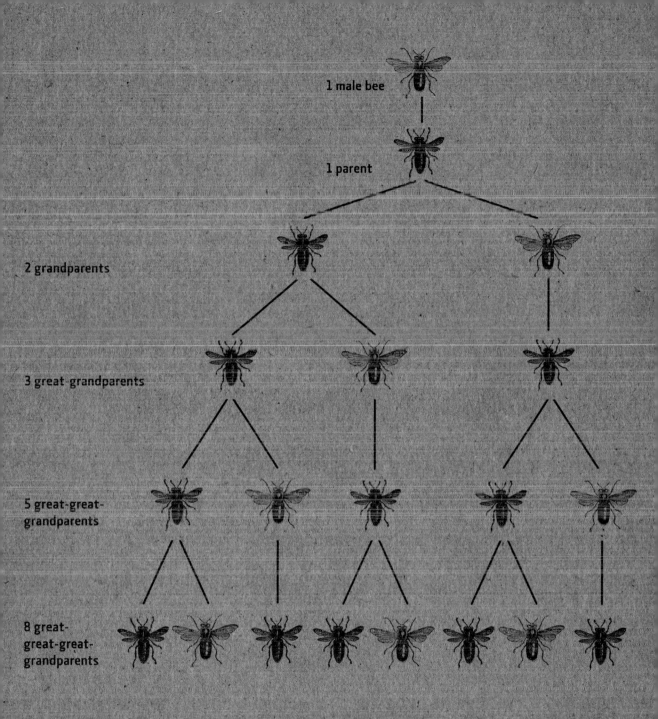

1 male bee

1 parent

2 grandparents

3 great-grandparents

5 great-great-
grandparents

8 great-
great-great-
grandparents

male bee   female bee

PASCAL'S TRIANGLE

the 30-second math

What comes next in this

sequence: (1 1), (1 2 1), (1 3 3 1), (1 4 6 4 1), ... ? This riddle is an important problem in algebra, known as "expanding brackets." Start with the expression $(1 + x)$ and multiply it by itself. This gives $(1 + x)^2 = 1 + 2x + 1x^2$. Multiplying three brackets gives $(1 + x)^3 = 1 + 3x + 3x^2 + 1x^3$. Four produces $(1 + x)^4 = 1 + 4x + 6x^2 + 4x^3 + 1x^4$. It is not the algebra that is difficult here, but the numbers. The next expression will look something like this: $(1 + x)^5 = 1 + ?x + ?x^2 + ?x^3 + ?x^4 + 1x^5$. But what are the right numbers to fill in here? Blaise Pascal wanted a way to find the answer quickly, and find it he did, in the rows of his famous triangle. It begins with a 1. Below that, there are two more 1s. Pascal's insight was that the process could be continued, with each number coming from the two above it, added together. (Earlier thinkers had come to similar conclusions, including the Indian thinker Pingala, more than a thousand years earlier.) This process is simple to do: just a little addition and no complicated algebra. Each row then gives the answer to a bracket–expanding problem. So to find $(1 + x)^5$, just read the numbers along the sixth row: 1,5,10,10,5,1.

3-SECOND SUM
Blaise Pascal's celebrated triangle not only contains many fascinating numerical patterns, it is also an essential tool in algebra.

3-MINUTE ADDITION
Pascal's triangle contains many fascinating patterns. The first diagonal is a row of 1s, and the second counts: 1, 2, 3, 4, ... But the third comprises what are known as the triangular numbers: 1, 3, 6, 10, 15, ... If you want to arrange balls into a triangle (at the start of a game of pool, for example), these are the numbers that work. The Fibonacci numbers are also hiding in the triangle, as the totals of successive "shallow diagonals"—see if you can find them!

RELATED THEORIES
See also
FIBONACCI NUMBERS
page 24
THE VARIABLE PLACEHOLDER
page 76
POLYNOMIAL EQUATIONS
page 80

3-SECOND BIOGRAPHIES
PINGALA
C. 200 BCE

ABU BEKR IBN MUHAMMAD IBN AL-HUSAYN AL-KARAJI
953–1029

YANG HUI
1238–1298

BLAISE PASCAL
1623–1662

ISAAC NEWTON
1643–1727

30-SECOND TEXT
Richard Elwes

Pascal's triangle contains numerous mathematical patterns and provides a neat solution to some algebraic problems.

```
                        1
                     1     1
                  1     2     1
               1     3     3     1
            1     4     6     4     1
         1     5    10    10     5     1
      1     6    15    20    15     6     1
   1     7    21    35    35    21     7     1
 1     8    28    56    70    56    28     8     1
1    9    36    84   126   126    84    36    9    1
1   10   45   120   210   252   210   120   45   10   1
1  11   55   165   330   462   462   330   165   55   11   1
1 12   66   220   495   792   924   792   495   220   66  12  1
```

June 19, 1623
Born in Clermont (now Clermont-Ferrand)

1631
Moved to Paris with his family

1639
Wrote *Essay on Conics*; family moved to Rouen

1642–1645
Constructed the Pascaline, a mechanical calculator

1647
Met Descartes and published *New Experiments Concerning Vacuums*

1650
Converted to Jansenism

1653
Returned to scientific study

1653
Published *Treatise on the Equilibrium of Liquids,* which explained his law of pressure

1654
Corresponded with Fermat

1655
Method for "Pascal's Triangle" printed; met Antoine Arthaud, leading Jansenist philosopher

1656–1657
Wrote *Lettres Provinciales*, in defence of Jansenism

1658
Wrote *Essay on the Cycloids*

1668
Began work on *Pensées*, a collection of philosophical and theological notes

August 19, 1662
Died in Paris

1670
Pensées published posthumously

1779
Essay on Conics published

BLAISE PASCAL

Suffering from chronic

migraines, insomnia, and dyspepsia, Pascal was in great pain for most of his short but productive life. Despite this, he became an outstanding mathematician, physicist, philosopher, and theologian, and worked (and quarreled) with the greatest minds of his time. Homeschooled and motherless from the age of six, Pascal was forbidden to study math by his father, so of course he did it in secret. When he was 12, his father relented and the young Pascal applied himself even harder, developing a calculating machine to help with his father's work as a tax collector. Called the "Pascaline," it was not the first mechanical calculator, and although 50 were made it was not a commercial success; but its design and the theory behind it were a great influence on Gottfried Leibniz.

Throughout his adult life, Pascal had regular spats with the philosopher Descartes over the existence (or not) of a vacuum. Descartes wrongly opined that there was no such thing, which led to Pascal's book on hydrostatics. He also found time to develop the idea of "Pascal's Triangle" (see pages 26–27), and establish the principles of probability theory in correspondence with Pierre de Fermat. We have inveterate gambler Chevalier de Méré to thank for this; he asked Pascal if he could work out how to divide up the pot if two players of equal abilities decided to quit the tables in mid-game. In 1646, Pascal's father fell ill and was nursed by Jansenist brothers from Port Royal monastery; Pascal and his sister Jacqueline were profoundly influenced by this and underwent religious conversion. Toward the end of his life, Pascal spent much of his time trying to reconcile faith and reason; his attempts are probably best exemplified by "Pascal's Wager," which appears in *Pensées*, a collection of philosophical considerations unfinished at the time of his death. The wager was concerned with the existence of God, and whether one should bet on it. Pascal set the odds in God's favour, reasoning that if He does exist, your place in heaven is secure, and if He doesn't, you have lost nothing.

NUMBER THEORY

the 30-second math

Number theory is the study of interesting properties that numbers possess. For example, choose any odd prime number and divide it by 4. The remainder will be either 1 or 3. It can be proven that if the remainder is 1, you can find two square numbers that add up to that prime. For example, dividing 73 by 4 gives 18 with a remainder of 1. After a short search, you can determine that $73 = 9 + 64 = 3^2 + 8^2$. On the other hand, a remainder of 3 means that no matter how hard you look, it is *impossible* to find two squares that add up to that prime (try 7 or 59). This begs the question: why? Mathematicians are never satisfied with discovering this kind of interesting behaviour—they want to prove that such properties are always true. Ancient Greek mathematicians began exploring properties of divisibility of integers, leading them to study prime numbers. They also enjoyed studying figurate numbers and their interrelationships. If you have a number of stones that can be arranged into an equilateral triangle, or a square, or a pentagon, and so forth, it is called figurate. Euclid even provided a formula for when any two squares add up to a third square. Pondering similar equations led Pierre de Fermat to conjecture what became his famous Last Theorem.

3-SECOND BIOGRAPHIES
PYTHAGORAS
c. 570–c. 490 BCE

EUCLID
fl. 300 BCE

PIERRE DE FERMAT
1601–1665

CARL FRIEDRICH GAUSS
1777–1855

G. H. HARDY
1877–1947

30-SECOND TEXT
David Perry

Figurate numbers are a branch of number theory—numbers that can expressed as a geometric arrangement.

3-SECOND SUM
Number theory is the discipline devoted to the study of properties and the behaviour of various classes of numbers.

3-MINUTE ADDITION
Carl Friedrich Gauss declared that mathematics was the queen of the sciences and that number theory was the queen of mathematics. G. H. Hardy echoed this sentiment some 70 years ago, relishing an area of mathematics that is only studied for the surprising beauty of the discovered truths, an area unsullied by practical application. When number theory later began to show unanticipated application to cryptology, few thought the beauty of the queen of mathematics was in any way diminished.

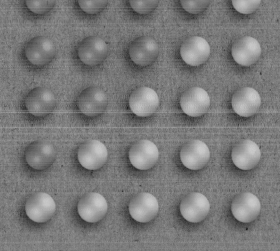

> Any square number is the sum of two triangular numbers—here 5^2 is the result of adding $10 + 15$.

> The addition of successive odd integers, starting with 1, gives you a square number: $8^2 = 64$.

1   3   5   7   9   11   13   15

MAKING NUMBERS WORK

algebraic expression Mathematical expression in which letters or other symbols are used to represent numbers. Algebraic expressions can also feature Arabic numerals and any of the signs of operation, such as + (addition), × (multiplication), √ (square root), and so on. No matter how complex the algebraic expression, it always represents a single value.

associative A property of an operation on numbers such that when an expression involves two or more occurrences of the operation it does not matter in which order the operations are performed. For example, multiplication of numbers is associative, since $(a \times b) \times c = a \times (b \times c)$.

Boolean logic (Boolean algebra) A form of algebra in which logical propositions are represented by algebraic equations in which "multiplication" and "addition" (and negatives) are replaced with "and" and "or" (and "no"), and where the numbers 0 and 1 represent "false" and "true" respectively. Boolean algebra played (and still plays) a significant role in the development of computer programming.

Cartesian coordinates Numbers that represent the position of a specific point on a graph or map by a grid-like positioning system. The coordinates are given by values representing the distance on both the horizontal (x) axis, and on the vertical (y) axis away from a reference point, usually the crossing point of the axes.

commutative A property of an operation on numbers such that when the order is reversed the answer is still the same. For example, multiplication of numbers is commutative since $3 \times 5 = 5 \times 3$.

differential equation An equation involving an unknown function and some of its derivatives. Differential equations are the primary tools used by scientists to model physical and mechanical processes in physics and engineering.

exponent The number of times by which another number, known as the base number, is to multiply itself. In the expression $4^3 = 64$, the exponent is 3 and the base is 4. The exponent is also known as the index or power.

expression A collection of numbers, and/or symbols, which together with any of the signs of operation, such as + (addition) or × (multiplication), determine a value.

function When applied to a quantity, known as the input, running a function results in another quantity, known as the output. A function is often written as $f(x)$. For example $f(x) = x^2$ is a function in that for every input value of x you get an output value of x^2, so that $f(5) - 25, f(9) - 81$, and so on. The collection of all inputs and outputs can be thought of as individual sets so that a function relates each element of the input set to another element of the output set.

monadology Gottfried Leibniz's metaphysical philosophy as expressed in his work *The Monadology* (1714). The philosophy is based around the concept of monads, simple substances Leibniz called "the elements of things," each of which is programmed to behave in a certain way.

multiplier The quantity by which a number, known as the multiplicand, is to be multiplied. In the expression $3 \times 9 = 27$, the multiplier is 3 and the multiplicand is 9.

number line The visual representation of all real numbers on a horizontal scale, with negative values running indefinitely to the left and positive to the right, divided by zero. Most number lines usually show the positive and negative integers spaced evenly apart.

quantum mechanics A branch of physics in which mathematical formulae play a central role in describing the motion and interaction of subatomic particles, including, for example, wave-particle duality.

real number Any number that expresses a quantity along a number line or continuum. Real numbers include all the rational numbers (that is, numbers that can be expressed as a ratio or fraction; including the positive and negative integers and decimals), the irrational numbers (those numbers that cannot be written as a vulgar fraction, such as $\sqrt{2}$), and the transcendental numbers (such as π).

variable A quantity that can change its numerical value. Variables are often expressed as letters such as x or y, and are often used as placeholders in expressions and equations such as $3x = 6$, in which 3 is the coefficient, x is the variant, and 6 is the constant.

ZERO

the 30-second math

Zero was used as a placeholder in numeral systems by several ancient peoples, including the Babylonians, Greeks (but only astronomers!), and Mayans. It was also used this way in India, where our modern system of numerals originated. In 628 CE, Brahmagupta wrote the first book that treats zero as a number rather than just a placeholder, giving rules for arithmetic using zero and negative numbers. Al-Khwarizmi introduced the Indian numeral system to the Islamic world in 820. Fibonacci introduced it to Europe in 1202 in *Liber Abaci*, popularizing the use of zero in Europe. Zero is the only real number that is neither positive nor negative, and any number that is not zero is called "non-zero." Zero is the additive identity, that is, $a + 0 = a$, where a is any real number, adding zero to it leaves it unchanged. Furthermore, $a \times 0 = 0$, and $0/a = 0$ for non-zero a. While one might think that a real number divided by zero is infinity, that doesn't make sense in a rigorous manner, so mathematicians just say division by zero is undefined. Because it is divisible by 2, 0 is an even number. However, if the exponent is 0 the answer is always 1, for example $a^0 = 1$ for any real number a other than 0. Some mathematicians prefer to count starting with 0 instead of 1.

3-SECOND BIOGRAPHIES
BRAHMAGUPTA
598–c. 670

ABU 'ABDALLAH MUHAMMAD
IBN MUSA AL-KHWARIZMI
c. 770–c. 850

LEONARDO FIBONACCI
c. 1170–c. 1250

30-SECOND TEXT
Robert Fathauer

3-SECOND SUM
Zero, the symbol for which is 0, is the absence of quantity. Synonyms for zero include nil, naught, zilch, zip, cipher, and goose egg.

3-MINUTE ADDITION
In Boolean logic 0 denotes false, and in electrical appliances 0 is shorthand notation for off. In physics, absolute zero is the theoretical minimum temperature. "Subzero" is used to mean negative numbers or quantities. To "zero" a device is to adjust it to zero value. And a "zero" is often used to mean an insignificant person or thing—hardly, for this very important and most versatile of our real numbers!

Much ado about nothing—zero is an integer in its own right.

INFINITY

the 30-second math

That the natural numbers are infinite (never ending) is easy to see. Declare any number to be the highest and you can always add one more. That there is an infinite number of numbers between 0 and 1 is also true, but a little trickier. The concept of infinity has fascinated mathematicians for millennia. The Greek stoic Zeno studied the idea through a series of paradoxes. His most famous posited that all motion is impossible, since to go from point A to point B, one must pass through an infinite number of in-between points, each taking a positive time to get from one to the next, and since an infinite number of positive numbers must add to infinity, one can go nowhere in finite time. We now know where he went wrong (an infinite number of positive numbers can have a finite sum!), but the thought provoked much study. Today the central idea behind calculus involves infinity. Average rates of change using an infinite sequence of increasingly small positive time intervals (we say "infinitesimally small") help define the instantaneous rate of change. This works much like a car's speedometer, which registers your speed—your distance traveled over a very small positive interval of time. Without infinity, maybe we really cannot go anywhere!

3-SECOND SUM
All good things must come to an end, but not in mathematics.

3-MINUTE ADDITION
Buzz Lightyear, the famous space hero in Pixar's *Toy Story* series, proudly proclaims "To infinity and beyond!" But like the end of the real number line and the horizon for intrepid sailors, no matter how far we travel, we are never any closer to it than when we started. Even the total number of subatomic particles in the universe, estimated at far less than 10^{100} (a googol), is no closer to infinity than 1 is. To get beyond infinity, one must first reach it. Even Zeno would have appreciated that.

RELATED THEORIES
See also
RATIONAL & IRRATIONAL NUMBERS
page 16
CALCULUS
page 50
THE CONTINUUM HYPOTHESIS
page 148

3-SECOND BIOGRAPHIES
ZENO OF ELEA
C. 490–C. 430 BCE

GEORG CANTOR
1845–1918

30-SECOND TEXT
Richard Brown

Will there ever be an end to all this? Not according to the mathematicians.

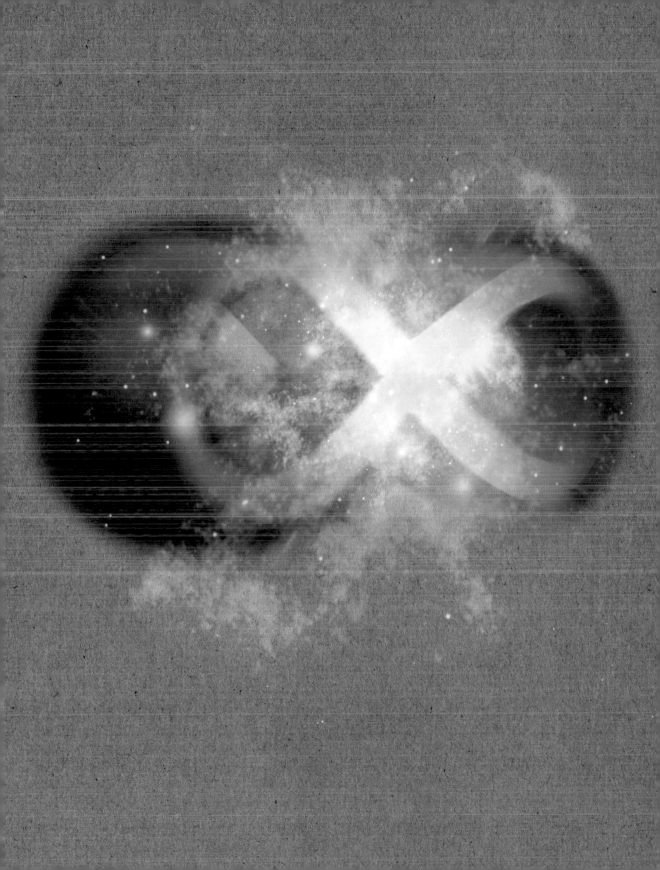

ADDITION & SUBTRACTION

the 30-second math

3-SECOND SUM
Addition is the combining of two or more numbers. Subtraction is taking the difference of two numbers.

3-MINUTE ADDITION
Infinitely many numbers may be added or subtracted in an infinite series. A series with a finite limit is said to be convergent. A simple example is the series $\frac{1}{2} + \frac{1}{4} + \frac{1}{8} + \frac{1}{16} + \ldots = 1$. To see this, imagine walking half way across a room, then half the remaining distance ($\frac{1}{4}$ of the total), then half the remaining distance ($\frac{1}{8}$) and so on. Some infinite series yield surprising results. For example, $1 - \frac{1}{3} + \frac{1}{5} - \frac{1}{7} + \frac{1}{9} - \frac{1}{11} + \frac{1}{13} - \frac{1}{15} \ldots = \pi/4$.

Ancient cultures such as the Egyptians and Babylonians were using addition and subtraction as early as 2000 BCE. The decimal numeral system used in India, which lent itself more readily to arithmetic operations, was adopted in Europe through Fibonacci's book *Liber Abaci*. Aryabhata and Brahmagupta made important contributions to Indian mathematics in the sixth and seventh centuries, and the + and − symbols first appeared in print in a book by Johannes Widmann published in 1489. In addition, the numbers being added are called the addends, and the result the sum. Carrying is performed when the sum of a column of addends is more than 9. Addition is commutative, meaning $a + b = b + a$, and associative, meaning $(a + b) + c = a + (b + c)$. Adding zero to a number results in that same number, making zero the additive identity, e.g., $a + 0 = a$. Subtraction is the inverse of addition. In subtraction, for example in $a - b$, a is the minuend and b the subtrahend. In contrast to addition, subtraction is neither commutative nor associative. Just as carrying is often required when adding a column of numbers, borrowing is often required when subtracting numbers. The symbol $\pm$, read "plus or minus," can be used to denote an uncertainty or a pair of values (for example the two roots of the quadratic equation).

RELATED THEORIES
See also
FRACTIONS & DECIMALS
page 14
COUNTING BASES
page 20
ZERO
page 36
MULTIPLICATION & DIVISION
page 42

3-SECOND BIOGRAPHIES
ARYABHATA
476–550
BRAHMAGUPTA
598–670
LEONARDO FIBONACCI
c. 1170–c. 1250
JOHANNES WIDMANN
c. 1462–c. 1498

30-SECOND TEXT
Robert Fathauer

The sum of all things—addition and subtraction have been part of daily life since ancient times.

2+35+12+514+147

74+568+44+235+89

142+23+25+28

+256+89+458+982+65

546+258+693+4

3+58+457+

MULTIPLICATION & DIVISION

the 30-second math

Multiplication and division were extremely challenging using early numeral systems that did not employ positional notation, such as Egyptian, Greek, and Roman numerals. The numeral and arithmetic system eventually adopted in Europe was developed in India, with important advances made in the sixth and seventh centuries. In the multiplication $a \times b = c$, a is the multiplier, b the multiplicand and c the product; a and b are also called factors. Notation for multiplication of two numbers a and b includes $a \times b$, $a \cdot b$, $(a)(b)$ and, favoured by mathematicians, simply ab. Similar to addition, carrying is necessary when the product of a column of digits is more than 9. In the example $a \times 1 = a$, 1 is the multiplicative identity. Multiplication is commutative, meaning $a \times b = b \times a$, and associative, meaning $(a \times b) \times c = a \times (b \times c)$. Division is neither. In the division $a \div b = c$, a is the dividend, b the divisor, and c the quotient. Mathematicians favour the notation a/b to $a \div b$. Long division is a division algorithm that displays the dividend (the amount to divide), divisor (number you divide by), and quotient (the answer) in a tableau. For mathematicians, division of any number by zero is undefined because it doesn't make sense in a rigorous manner.

3-SECOND SUM
Multiplication is repeated addition of a first number a specified second number of times. Division is determining how many times one quantity is contained in another.

3-MINUTE ADDITION
Using logarithms, multiplication and division can be performed using addition and subtraction, respectively. This is made possible by the fact that multiplying or dividing numbers expressed as powers of a common base can be accomplished by adding or subtracting the exponents. Before the advent of desk and handheld calculators, slide rules marked with logarithmic axes were commonly employed to facilitate arithmetic calculation.

RELATED THEORIES
See also
FRACTIONS & DECIMALS
page 14
NUMBER THEORY
page 30
ADDITION & SUBTRACTION
page 40
EXPONENTIALS & LOGARITHMS
page 44

3-SECOND BIOGRAPHIES
ARYABHATA
476–550

BRAHMAGUPTA
598–670

LEONARDO FIBONACCI
C. 1170–C. 1250

30-SECOND TEXT
Robert Fathauer

Multiplication takes one number and repeats it by a second number of times. Division is the opposite, splitting one number into equal portions.

x	1	2	3	4	5	6	7	8	9	10
1	1	2	3	4	5	6	7	8	9	10
2	2	4	6	8	10	12	14	16	18	20
3	3	6	9	12	15	18	21	24	27	30
4	4	8	12	16	20	24	28	32	36	40
5	5	10	15	20	25	30	35	40	45	50
6	6	12	18	24	30	36	42	48	54	60
7	7	14	21	28	35	42	49	56	63	70
8	8	16	24	32	40	48	56	64	72	80
9	9	18	27	36	45	54	63	72	81	90
10	10	20	30	40	50	60	70	80	90	100

$$7\overline{)42} = 6$$

EXPONENTIALS & LOGARITHMS

the 30-second math

If I add £1 to my piggy bank every week and track the amount I've saved, I will chart an amount that grows linearly (at a constant rate). If I add £1 every week to a bank account that gains interest, the amount will grow exponentially (at a rate that is increasing along with the amount itself, as we start generating interest on previously earned interest, which has a cascading snowball effect). A generous bank might give me a 100% interest rate, meaning I would earn £1 interest on the original £1 I invested, giving me £2 after one year. If I never added money but just left that amount to continue to accrue interest, it would double every year, giving me £8 after three years, because $2 \times 2 \times 2 = 2^3 = 8$. After four years, I would have £16, and so on. In the expression $2^3 = 8$, we call the constant multiplier 2 the base; the exponent 3 is the number of times we multiply the base by itself. It is natural to want to reverse this calculation. What if I know the interest rate but want to know how many years it will take before £1 becomes £8? A logarithm reverses the exponentiation, and we write $\log_2 8 = 3$. In general, the function $\log_2$ tells me what exponent to raise 2 by to get x. In the bank example, with my money doubling every year, it tells me how many years it will take to earn £x.

RELATED THEORIES
See also
RATIONAL & IRRATIONAL NUMBERS
page 16
MULTIPLICATION & DIVISION
page 42
FUNCTIONS
page 46

3-SECOND SUM
Exponentiation is a shorthand notation for repeated multiplication. A logarithm is to exponentiation as division is to multiplication—a mathematical way to undo it.

3-MINUTE ADDITION
The mathematician John Napier first used the term logarithm to denote the inverse of exponentiation, and in the 16th century produced tables of values to calculate logarithms. You have likely seen buttons on your calculator for $\log_{10}(x)$ (the logarithm for base 10) and $\ln(x)$, referred to as the 'natural logarithm'. The base for this logarithm is a number between 2 and 3 called e, a special number, like π, frequently seen in formulas in physics, biology, and economics.

3-SECOND BIOGRAPHIES
JOHN NAPIER
1550–1617
LEONARD EULER
1707–1783

30-SECOND TEXT
David Perry

Whereas logarithmic growth tapers off drastically, exponential growth is explosive.

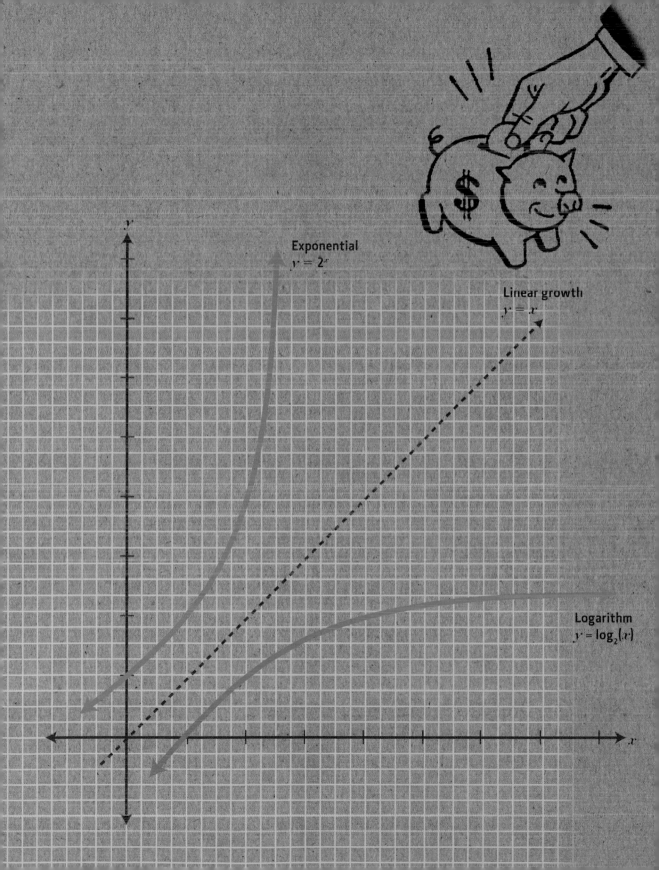

Exponential
$y = 2^x$

Linear growth
$y = x$

Logarithm
$y = \log_2(x)$

FUNCTIONS

the 30-second math

Examples of functions are

found very early in recorded history, but the modern concept of the mathematical function appears much later. In its most basic form, a function is a relationship that creates a single output value for a single input value. The symbol $f(x)$ is used to denote a function of the variable x. For example, $f(x) = x^2$ is a function for which an output value of 9 (3^2) is obtained for an input value of 3. In the 14th century, the work of Oresme included ideas of dependent and independent variables. Galileo constructed formulae that mapped one set of points to another, and Descartes introduced the concept of constructing a curve using an algebraic expression. The term "function" was coined by Leibniz in the late 17th century. The set of all inputs of a function is called the domain, while the set of all outputs is called the image or range. Functions of a single variable (or argument) are often plotted using Cartesian coordinates, where x is the abscissa (horizontal axis), and $f(x)$ is the ordinate (vertical axis). For example, for $f(x) = 2x + 3$ a graph of $f(x)$ would show a line made up of all ordered pairs (x, y) that satisfy this equation. These include $(1,5)$, since $5 = 2 \times 1 + 3$ and $(2,7)$ since $7 = 2 \times 2 + 3$. Functions of the two variables can be plotted with $f(x, y)$ as the vertical axis and the x–y plane lying horizontal.

RELATED THEORIES
See also
EXPONENTIALS
& LOGARITHMS
page 44

THE EQUATION
page 78

TRIGONOMETRY
page 102

GRAPHS
page 108

3-SECOND BIOGRAPHIES
NICOLE D'ORESME
c. 1320–1382

RENÉ DESCARTES
1596–1650

GOTTFRIED LEIBNIZ
1646–1716

30-SECOND TEXT
Robert Fathauer

3-SECOND SUM
A mathematical function is a relation that associates each element of a set with an element of another set.

3-MINUTE ADDITION
The concept of a function is widely employed in the physical sciences and engineering, in which case the function and its arguments usually correspond to measurable physical quantities like temperature, volume, and gravitational attraction. Functions are also commonly used in economics and business, where the variables could be demand, time, interest, profit, and so on. Indeed, studying the functional relationships between two or more entities is at the core of understanding the mathematical processes of nature and business. Works for understanding people, also, no?

When any value of x is plugged into the equation $-1.7x^3 - 5x^2 - 0.3x + 1$, the result it yields can be plotted on a graph, giving a visual representation of the function.

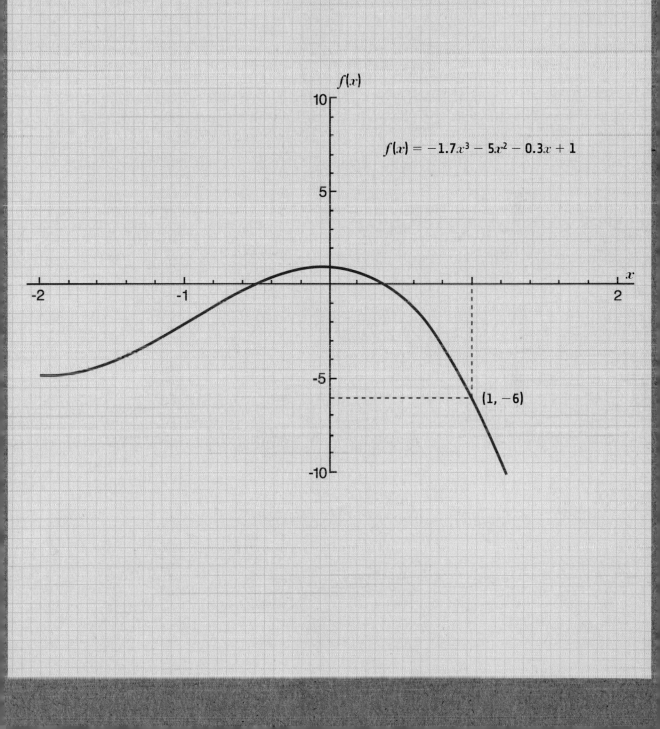

> This plot shows the value of $f(x)$ in the range shown from –2 to approximately 1.2. For example, at $x = 1$ the result is –6. So one of the points making up the curve is described by the coordinates (1,–6).

$f(x) = -1.7x^3 - 5x^2 - 0.3x + 1$

$f(x)$

10

5

x

-2

-1

2

-5

(1, −6)

-10

July 1, 1646
Born in Leipzig

1662
Completed BA in
Philosophy at
University of Leipzig

1664
Gained Masters degree
in Philosophy

1665
Gained BA in Law

1673
Elected as a member of
the Royal Society and
appointed counsellor by
the Duke of Brunswick

November 1675
Achieved breakthrough
in infinitesimal calculus

1677
Appointed Privy
Counsellor of Justice to
the House of Brunswick

1684
Published his notes
on calculus

1686
Published *Discourse
on Metaphysics*

1710
Theodicee published

1711
Accused of plagiarism

1712–1714
Wrote *Monadology*

November 14, 1716
Died in Hanover

GOTTFRIED LEIBNIZ

A gifted polymath of the late 17th and early 18th centuries, whose work is, in the main, written down in short treatises, notes, papers in learned journals and correspondence, Leibniz suffered the curse of the early adopter. This may reflect the sheer breadth of his intellectual application. Many of Leibniz's ideas foreshadow modern thought and theory in the fields of physics, technology, biology, medicine, geology, psychology, linguistics, politics, law, theology, history, philosophy, and mathematics. He improved on Pascal's calculating machine (anticipating the work of Babbage and Lovelace); developed binary theory which underpins modern digital technology; developed what we now know as Boolean algebra and symbolic logic; and outlined the concept of feedback that inspired Norbert Wiener.

An academic wunderkind and son of a university professor, Leibniz was fluent in Latin at the age of 12 and took his first degree at 16. The holder of degrees in mathematics, philosophy, and law, he later eschewed academia and spent most of his life working under the patronage of the House of Brunswick, living and working in Leipzig, Paris, London, Vienna, and Hanover, meeting and corresponding with leading scientists and philosophers of his day. Probably his best-known philosophical theory is monadology (monads being the smallest indivisible unit of philosophical thought). Sadly however, for such an intellectual powerhouse, due to a bitter controversy he was not recognized at his death, despite his royal and intellectual connections, and his grave went unmarked for 50 years. The controversy, between Leibniz and Newton over who invented calculus, sprang up in 1711 and has never gone away. Leibniz knew Newton, was a fellow member of the Royal Society, and had been in London at the same time that Newton was developing calculus; when Leibniz brought out his own version, most mathematicians sided with Newton, and Leibniz was vilified. Whether or not he stole the idea and presented it as his own, or whether they both came to the same conclusion working in ignorance of each other, may never be known and today they are both credited with the invention.

CALCULUS

the 30-second maths

Many branches of science study objects that move and change over time. As a ball rolls down a hill, for instance, its position changes. The rate of change of position is the ball's speed. But, of course, that may change too. The rate of change of speed is called acceleration. The question is, if you have a mathematical formula describing the position of the ball, can you then calculate its speed and acceleration? The geometrical problem is to start with a curved line in the plane, and determine how steep it is at any given point. If the curve is a graph of a ball's position against time, then its steepness represents the ball's speed. This had been understood since the time of Archimedes, but only approximate methods were originally known for calculating the all-important steepness of the curve. In the late 17th century, Isaac Newton and Gottfried Leibniz separately developed calculus, a beautiful set of rules for describing the steepness of graphs and related ideas. The subject has two branches. Starting with a curve, differential calculus will tell you its steepness, while integral calculus describes the area trapped underneath it. Unexpectedly, these are opposite procedures, a fact known as the fundamental theorem of calculus.

RELATED THEORIES
See also
THE EQUATION
page 78
GRAPHS
page 108

3-SECOND BIOGRAPHIES
ARCHIMEDES
C.287–212 BCE

ISAAC NEWTON
1643–1727

GOTTFRIED LEIBNIZ
1646–1716

AUGUSTIN-LOUIS CAUCHY
1789–1857

KARL WEIERSTRASS
1815–1897

30-SECOND TEXT
Richard Elwes

From the position of a traveling ball, calculus can tell us its speed and acceleration. When applied to a hill, calculus produces the tangent plane that determines the hill's steepness.

3-SECOND SUM
Calculus is a branch of mathematics that describes how systems and other mathematical constructions change across time and space.

3-MINUTE ADDITION
The discovery of calculus by Newton and Leibniz is one of the most important moments in mathematical history. From climate modeling and economics to quantum mechanics and relativity theory, a huge range of applications of mathematics to the physical world are expressed in terms of "differential equations" and studied via calculus. Solving these sorts of equations is therefore one of the biggest technical challenges for today's scientists and mathematicians.

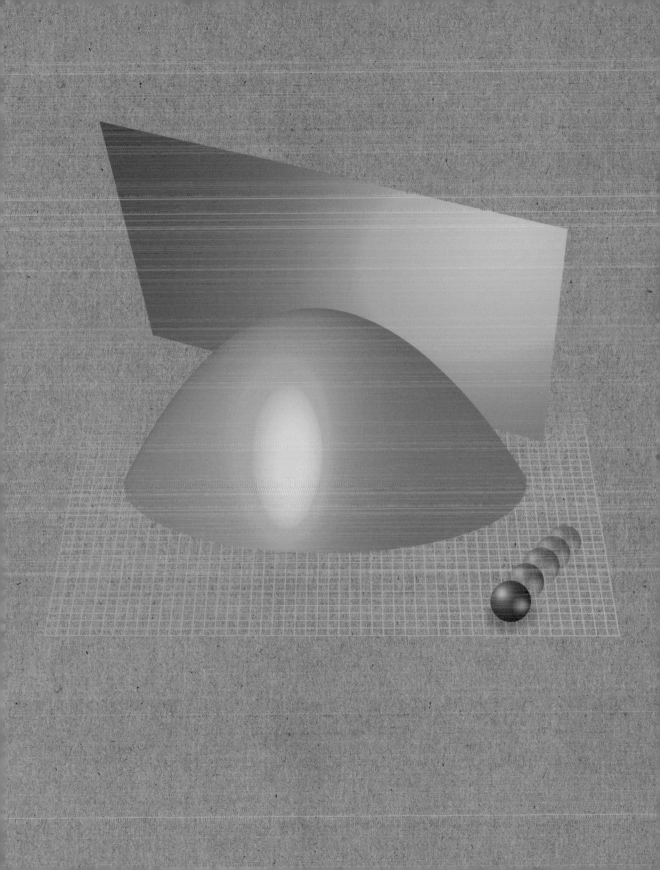

CHANCE IS A FINE THING

bell curve In probability theory, the name given to describe the shape of a smooth graph representing a standard normal distribution. The peak of the curve represents the mean, down from which two sloping sides, equal in shape and representing all possible variations, drop quickly before flattening out.

binary sequence In computer science, a long string of "0s" and "1s" that represent "off" and "on" respectively. Binary sequences essentially provide instructions for a computer.

central limit theorem In probability theory, the central limit theorem states that if an equally random variable, such as throwing a dice, is performed a sufficient number of times, the mean will tend towards normal; and the results, if plotted on a graph, will describe a bell curve.

equilibrium In game theory, equilibrium describes the point in a game at which all players are employing strategies that ensure no player has a greater chance of winning.

false positive Name given to an error in, for example, a medical trial. False positives occur due to the inaccuracy of the testing procedure resulting in a positive reading or result when in actual fact the true reading or result should be negative. Due to the occurrence of false positives in many testing environments it's impossible accurately to determine the probability of something or someone testing positive until there is sufficient data to calculate prior probability. See *prior probability*; *true positive*.

frequency The number of times a specific event occurs during a set period of time or over a larger set of trials of an experiment. The greater the number of occurrences, the higher or greater the frequency.

odds Odds express the likelihood of something happening by measuring the ways it would happen against the ways it wouldn't. If the probability of an event happening is p, and its probability of not happening is $1 - p$, then the "odds" in favour of it happening are $p/(1 - p)$. The "odds" against it happening are $(1 - p)/p$. For example, the probability of rolling a 4 with a standard die is 1/6. The probability against rolling a 4 is 5/6. The "odds" in favour of rolling a 4 are then $(1/6)/(5/6)$, or 1/5. Expressed the usual way, we would say the odds of rolling a 4 are 1:5 in favour. The "odds" against rolling a 4 are 5:1 (against). This means that there are "five ways to lose for every one way to win."

prior probability In statistics, the probability of an event that is set before new data or evidence is tested to calculate other probabilities. Prior probability plays a crucial role in Bayes' theorem of probability.

probability Probability is a way of expressing the likelihood of a specific event occurring by comparing it against all possible outcomes. It is the ratio of the number of desirable outcomes to the number of possible outcomes, which is then written as a number between 0 (zero likelihood) and 1 (certainty). For example, when picking a card from a full deck, the probability of choosing a heart is 13/52 or 1/4. So the probability of choosing a heart is 0.25.

true positive An accurate positive result given during, for example, a medical trial. True positives differ from false positives in that, whereas a true positive is a truly accurate positive result, a false positive is an inaccurate positive result that occurs due to an inaccuracy or failure in the testing procedure. See *false positive*.

GAME THEORY

the 30-second math

For millennia people have enjoyed games of strategy, from tic-tac-toe to chess and checkers. Some are easier than others. In tic-tac-toe for instance, it is fairly easy to formulate a good strategy. With a little practice, you should never lose. Game theory is the mathematical study of such strategies. Take a game like "rock, paper, scissors." What is the best strategy for winning here? If you decide to play scissors more often than paper or rock, then your opponent can exploit this by increasing the number of times she plays rock. Unless you can find a pattern in your opponent's behaviour, however, the best long-term strategy is to pick from the three options at random each time. When playing this way, you will win, lose, and draw equally often. This is what is known as an "equilibrium" of the game, since if both players are using this strategy, there is no way either of them can increase their number of wins by changing tactic. A centerpiece of game theory is the celebrated fact, proved by John von Neumann and expanded by John Nash, that a huge variety of games are guaranteed to have equilibria.

RELATED THEORIES
See also
THE LAW OF LARGE NUMBERS
page 62
THE GAMBLER'S FALLACY—
LAW OF AVERAGES
page 64
THE GAMBLER'S FALLACY—
DOUBLING UP
page 66
BAYES' THEOREM
page 70

3-SECOND BIOGRAPHIES
JOHN VON NEUMANN
1903–1957

CLAUDE SHANNON
1916–2001

JOHN NASH
1928–

JOHN CONWAY
1937–

30-SECOND TEXT
Richard Elwes

3-SECOND SUM
The strategies used in games such as chess can be mathematically analyzed, and appear in a wide range of scientific subjects.

3-MINUTE ADDITION
Game theory has moved beyond the study of games, with applications from political science to artificial intelligence. But games still pose challenges. In 2007, Canadian professor Jonathan Schaeffer and colleagues developed an infallible strategy for checkers. Their program will never lose. While computers can beat humans at chess, a perfect strategy like this remains a distant dream. The obstacle is the sheer number of ways a game of chess can develop, far outnumbering the number of atoms in the universe.

Rock, paper, scissors—do you have a strategy? Mathematicians do.

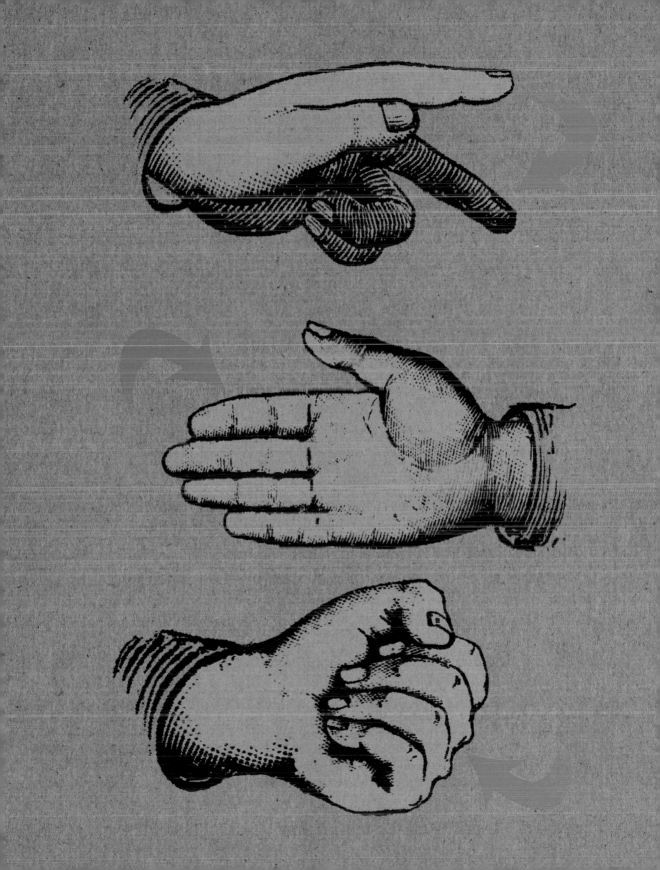

CALCULATING THE ODDS

the 30-second math

If you roll a die, the odds of getting a 6 are "5 to 1" against. This means that there are six outcomes in total, all equally likely, of which five are unsuccessful and one is a success. A mathematician would express the same thing via a fraction, by saying the "probability" of getting a 6 is 1/6; one successful outcome out of the six total possibilities. Similarly, the odds of pulling the ace of spades from a standard deck of cards are 51 to 1 against, or 1/52. So long as all outcomes are equally likely (so that the dice or cards are unbiased), these odds can be calculated by counting successful and unsuccessful outcomes. The science of probability assigns numbers to events to describe their likelihood of happening. These numbers always sit between 0 and 1, with 0 corresponding to impossible events, and 1 to certainties. Unlikely events have low probabilities: if you flip a fair coin ten times, the chance of getting ten heads is 1/1024 (1023 to 1 against). On the other hand, likely events have high probabilities (and good odds): if you pick a card from a deck, the chance of avoiding the ace of spades is 51/52 (or 1 to 51). A safe bet, no?

RELATED THEORIES
See also
THE LAW OF LARGE NUMBERS
page 62
THE GAMBLER'S FALLACY—
LAW OF AVERAGES
page 64
RANDOMNESS
page 68
BAYES' THEOREM
page 70

3-SECOND BIOGRAPHIES
PIERRE DE FERMAT
1601–1665

BLAISE PASCAL
1623–1662

CHRISTIAAN HUYGENS
1629–1695

ANDREY KOLMOGOROV
1903–1987

30-SECOND TEXT
Richard Elwes

When you roll a die, the likelihood of rolling an odd number is 3/6, so the odds are 1 to 1 or "even money" —three ways to lose and three ways to win.

3-SECOND SUM
Likely and unlikely events can be measured on a scale, in the language either of bookmakers' odds or of mathematicians' probabilities.

3-MINUTE ADDITION
Bookmakers offer better odds (and more money) on events that are very unlikely to happen. That is why we use the word "against." Long odds mean that the event is unlikely; be careful betting on a 40 to 1 horse, no one else thinks he's a winner. It's possible, but his probability of winning is 1/41. On the other hand, short odds like 2 to 3 against help define the favourite (3/5 probability of winning). The payout will be small, but at least you are "playing the odds."

1501
Born 24 September in Pavia, Italy

1520
Enrolled at the University of Pavia

1525
Achieved Doctorate in Medicine from the University of Pavia; applied to College of Physicians in Milan, but rejected until 1539

1526
Wrote *Liber de ludo aleae* (*On Casting the Die*), published posthumously in 1663

1536
Wrote *De malo recentiorum medicorum usu libellus* (on medicine)

1539
Wrote *Practica arithmetice et mensurandi singularis* (on mathematics)

1545
Wrote *Artis magnae, sive de regulis algebraicis* (also known as *Ars magna*)

1545
Cast and published the horoscope of Jesus Christ

1550
Invented the Cardan grille, a cryptographic tool

1570
Accused of heresy

1570
Wrote *Opus novum de proportionibus* (on mechanics)

1576
Died 21 September in Rome

1576
De vita propria (autobiography) published on his death

GIROLAMO CARDANO

Doctor, mathematician,

geologist, natural scientist, alchemist, astrologer, astronomer, and inventor, Cardano was the incarnation of Renaissance man (the exception to his genius being the arts)—a dark mirror to Leonardo da Vinci, a family friend with whom he sometimes collaborated (detractors say plagiarized). Both were the illegitimate sons of lawyers, both were men of exceptional talent; Leonardo went on to fame and glory, but Cardano's unpleasant personality and hypercritical manner nullified his gift and, despite being greatly sought after for his intellect, he managed to make himself loathed almost everywhere he went.

Medicine was his first career; he was an excellent clinician, consulted by the great, yet full of open contempt for his colleagues; lacking a bedside manner, or manners, he failed to make his medical practice at Sacco flourish, although he was to be compared later with Vesalius and became Professor of Medicine at the University of Pavia, his alma mater.

He turned his mind to mathematics, which he had studied with his father, and produced two books, one of which, *Ars magna* (1545) is a key Renaissance text that tackles the solution of cubic and quartic equations (see pages 80–81). Again he courted controversy; he had extracted the proof of cubic equations from Niccolò Tartaglia, who told Cardano on the promise that he would not publish for six years. However, discovering that Tartaglia had been rather economical with the truth, Cardano went ahead, published, and was damned by Tartaglia and his many enemies.

Disaster struck in 1560, when Cardano's revived medical career was blooming. His eldest son murdered his adulterous wife, and was tried and executed. His death devastated Cardano and ruined him professionally; he moved to Rome, stripped of his professorships, and was briefly imprisoned for heresy for casting a horoscope for Jesus Christ. Throughout his controversial career, Cardano had been addicted to gambling; he was very good at it, and wrote a book, *Liber de ludo aleae* (*On Casting the Die*), the first to look at probability based on what comes up when the die is rolled—in mathematical terms. Some purists sneer, but it is a great favorite with gamblers and casino owners, mainly because it contains a very good section on how to cheat. After a long, prolific but chaotic life, Cardano died on September 21, 1576. It is said that he predicted his death to the hour. It is also said that he committed suicide at the appointed time, so that he would not be proved wrong.

THE LAW OF LARGE NUMBERS

the 30-second math

3-SECOND SUM
Given enough trials, the frequency of a chance event will be very close to the probability of it occurring.

3-MINUTE ADDITION
The first significant step in demonstrating a relation between probability and frequency was taken by Jacob Bernoulli in 1713. This was reinforced by the work of Irénée-Jules Bienaymé and Pafnuty Chebychev 150 years later, and the icing on the cake, giving complete confidence that estimates will eventually be as good as we would like, came from Émile Borel in 1909.

Take any experiment with chance outcomes—such as throwing a ball through the top of a basket-ball hoop or tossing a coin —that is repeatable as often as you like under the same conditions. The probability of flipping ten heads in a row is small, but it is possible. If we flip this coin for ever, unlikely events like this one will occur from time to time. But in the long run, the percentage of occurrence of, say, heads will home in on its probability of occurrence. This is the law of large numbers—it is the principle that, in the long run, the probability of an event occurring determines its eventual frequency of occurring. The law of large numbers isn't restricted only to chance events. Say you want to know the average height of women living in Britain. In studying large populations, the larger the sample size, the better the average of the sample represents the average of the population. The precision of your estimate of an average increases only with the square root of the sample size. And for a good estimate, you need a larger sample when what you are measuring has higher variability. But this law assures us that, with enough data, we can always get as good an estimate as we need.

RELATED THEORIES
See also
THE GAMBLER'S FALLACY—
LAW OF AVERAGES
page 64

3-SECOND BIOGRAPHIES
JACOB BERNOULLI
1654–1705

IRÉNÉE-JULES BIENAYMÉ
1796–1878

PAFNUTY CHEBYCHEV
1821–1894

ÉMILE BOREL
1871–1956

30-SECOND TEXT
John Haigh

What are the chances of shooting three out of ten hoops over a period of time? In the long run they are pretty much the same.

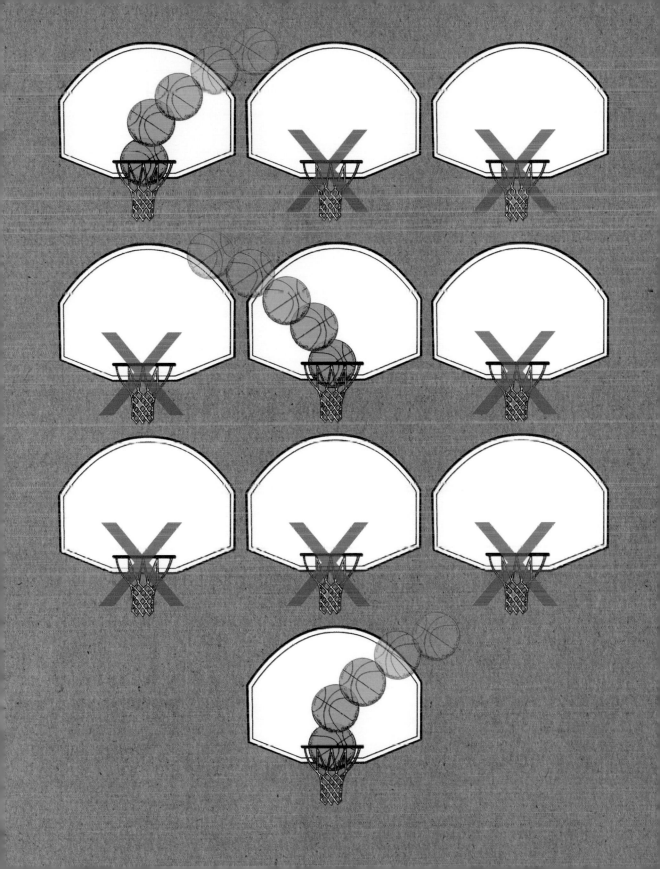

THE GAMBLER'S FALLACY—LAW OF AVERAGES

the 30-second math

3-SECOND SUM

In games of chance, using previous performance to bet on future behavior is definitely a losing strategy.

3-MINUTE ADDITION

Coins, dice, and roulette wheels all have outcomes that are equally likely at each trial. And unlikely events do occur: ten heads in a row, 12 consecutive rolls of "7," no number above "30" among 20 spins, and so on. There are so many "rare" things that might happen that some of them must occur ("rare events happen often!"). But they can in no way affect future performance or our predictions of it.

When a series of ten coin flips all show heads, it is tempting to argue that tails is more likely next time. People say, "By the law of averages that heads and tails are equally likely, tails must start to catch up." Nonsense: with a fair coin, no matter what the previous outcomes have been, the chances of heads or tails next time remain fixed at 50% heads, 50% tails. Similarly with roulette and lotteries: the fact that zero has not come up for 100 spins does not increase the chance it will come up next time. In Italy, the number 53 failed to appear in the lottery for more than two years, apparently resulting in numerous bankruptcies and suicides. Coins, roulette wheels and lottery balls are inanimate objects with no ability to remember previous outcomes and adjust their frequency. Frequencies will settle down to their different probabilities, in the long run—which may take a very long time indeed! Any genuine "law of averages" is strictly a paraphrase of the law of large numbers, and cannot be used to claim that past results will influence the immediate future.

RELATED THEORIES

See also
THE LAW OF LARGE NUMBERS
page 62
THE GAMBLER'S FALLACY—
DOUBLING UP
page 66

3-SECOND BIOGRAPHY
GIROLAMO CARDANO
1501–1576

30-SECOND TEXT
John Haigh

Each time you flip a coin, the chances of getting heads or tails always remains the same—even if you flip several heads or tails in a row.

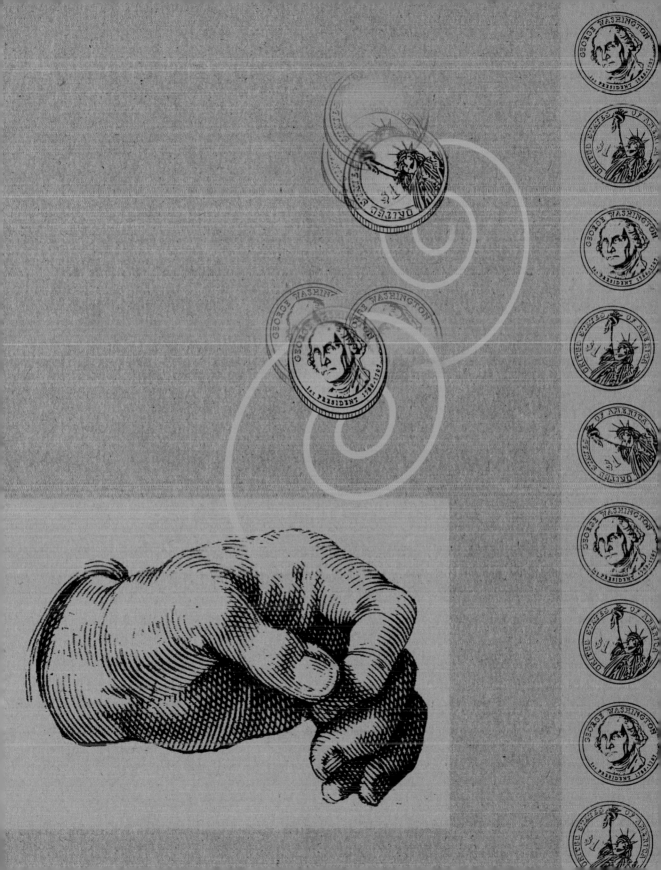

THE GAMBLER'S FALLACY— DOUBLING UP

the 30-second math

RELATED THEORIES
See also
THE LAW OF LARGE NUMBERS
page 62
THE GAMBLERS'S FALLACY—
LAW OF AVERAGES
page 64

3-SECOND BIOGRAPHY
GIROLAMO CARDANO
1501–1576

30-SECOND TEXT
John Haigh

3-SECOND SUM
In roulette, doubling your stakes after each loss on red/black bets is a losing, not a winning, strategy.

3-MINUTE ADDITION
American wheels have an additional "double-zero," but the payout odds are the same. In either case, the casino's advantage on any bet is small, but real. There is no way of combining different bets on one spin, or combining bets on different spins, to overcome this advantage. If the roulette wheel is in pristine condition, with all outcomes random every time, and a maximum stake is imposed, a gambler will lose, in the long run.

A European roulette wheel has 37 slots, comprising 18 red, 18 black, and one green (0). Bets on red or black pay out at even money. A gambler resolves always to bet on red, and to double up his bet after a loss. Since the chance of red is non-zero at any spin, it is inevitable that red turns up sometime; maybe the first red occurs on the fourth attempt: he has losses of size 1, 2, and 4 (total 7), then a profit of 8, resulting in a net profit of 1 unit. This 1-unit profit always arises, no matter how long it takes for the first red to arise. The gambler argues that he inevitably wins 1 unit whenever red appears. Unfortunately for the gambler, this is false. All casinos impose a maximum stake, usually around 100 times the minimum. So after seven losses of size 1, 2, 4, 8, 16, 32, 64 (total 127), casino rules prevent the required stake of 128 units, even if the gambler possesses the necessary capital to make the bet! The gambler may use this system and win 1 unit several times, but it is inevitable that, at some stage, the size of bet his system demands is not permitted; his losses will more than wipe out his gains.

Don't bet on doubling your stakes—it's a losing game.

RANDOMNESS

the 30-second math

Imagine two long sequences of heads (H) and tails (T), each beginning HHTHTH … One is truly random, the result of repeatedly tossing an unbiased coin. The other is not; it is carefully chosen by a human being. Is there any way of telling which is which? One simple test says that, in the long term, heads and tails should appear equally often in a random sequence. But this alone is not enough. It should also be that every pair of results (HH, HT, TH and TT) should, on average, appear equally often as every other. The same is true of every triple, quadruple or longer sequence. But all of these are not enough, since it is still possible to meet these conditions artificially. The simplest sequence runs HHHHHH … This is obviously non-random. But there is something else: it can be easily compressed. The phrase "one million heads" describes this sequence very succinctly, and allows anyone to communicate and recreate it with perfect accuracy. Truly random sequences cannot be compressed at all. The only way to communicate a random sequence to someone else is by writing it out in full. It is a deep, recent discovery that randomness and incompressibility are essentially the same thing.

3-SECOND SUM
Randomness is central to science, but very difficult to detect mathematically.

3-MINUTE ADDITION
The Internet runs on binary sequences: long strings of 0s and 1s that computers can translate into all the programs and files we wish to use. For maximum efficiency, these strings should be compressed as much as possible, by using file-compression software. When a string has been compressed, by stripping out any predictable or repetitive patterns, it becomes indistinguishable from a purely random sequence. Perfectly compressed information is therefore mathematically identical to randomness.

RELATED THEORIES
See also
THE LAW OF LARGE NUMBERS
page 62
BAYES' THEOREM
page 70
ALGORITHMS
page 84
GÖDEL'S INCOMPLETENESS
THEOREM
page 144

3-SECOND BIOGRAPHIES
EMILE BOREL
1871–1956
ANDREY KOLMOGOROV
1903–1987
RAY SOLOMONOFF
1926–2009
GREGORY CHAITIN
1947–
LEONID LEVIN
1948–

30-SECOND TEXT
Richard Elwes

Which sequence is random? Even the mathematicians can't tell.

BAYES' THEOREM

the 30-second math

3-SECOND BIOGRAPHY
THOMAS BAYES
c. 1702–1761

30-SECOND TEXT
Jamie Pommersheim

Suppose that a test for a certain disease is 90% accurate. Now suppose that a randomly chosen person, Bob, tests positive. What is the probability that Bob actually has the disease? It turns out that you can't answer this question! You need one additional piece of information, namely how common the disease is. That is, you need to know the prior probability that a randomly chosen person has the disease. Let's suppose that 1% of the population has the disease. Bayes' theorem tells us how to find the probability of having the disease given a positive test. In a group of 1,000 people, on average 10 have the disease (1%) and 9 of these will test positive ("true positives"). The remaining 990 do not have the disease, and 10% of these, or 99, will still test positive ("false positives"). The false positives outnumber the true positives by 99 to 9, so the odds are 11:1 against Bob having the disease. An unlikely event remains unlikely even in spite of the evidence provided by the accurate test!

3-SECOND SUM
Bayes' theorem helps you find the likelihood of an event given all the evidence, but only if you know the prior probability of the event.

3-MINUTE ADDITION
Bayes' theorem is named after the Reverend Thomas Bayes, a Presbyterian minister who lived in 18th-century England. His work on the subject was not published until several years after his death. Bayes' theorem raises philosophical questions about the very nature of probability. In particular, the appearance of prior probabilities in Bayes' theorem suggests that you cannot meaningfully assign probabilities to events without first using repeated trials to determine the frequency of the event.

The odds of an event happening is the ratio of the number of true positives (9) to the number of false positives (99).

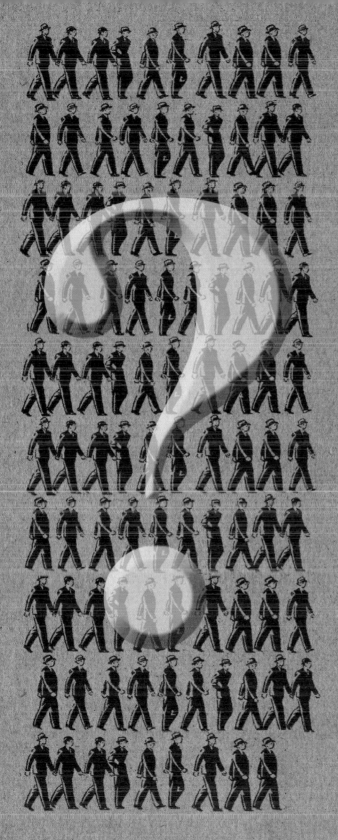

ALGEBRA & ABSTRACTION

algebraic geometry The branch of mathematics that combines geometry with algebra; it involves the study of geometric shapes that are created from the graphs of solutions to algebraic polynomial equations.

associative A property of an operation on numbers such that when an expression involves two or more occurrences of the operation it does not matter in which order the operations are performed. For example, multiplication of numbers is associative, since $(a \times b) \times c = a \times (b \times c)$.

coefficient A number that is used to multiply a variable; in the expression $4x = 8$, 4 is the coefficient, x is the variable. Although usually numbers, coefficients can be represented by symbols such as a. Coefficients that have no variables are called constant coefficients or constant terms.

commutative A property of an operation on numbers such that when the order is reversed the answer is still the same. For example, multiplication of numbers is commutative since $3 \times 5 = 5 \times 3$.

constant A number, letter, or symbol on its own that represents a fixed value. For example, in the equation $3x - 8 = 4$, 3 is the coefficient, and x is the variable, while 8 and 4 are the constants.

differential equation An equation involving an unknown function and some of its derivatives. Differential equations are the primary tools used by scientists to model physical and mechanical processes in physics and engineering.

exponent The number of times by which another number, known as the base number, is to multiply itself. In the expression $4^3 = 64$, the exponent is 3 and the base is 4. The exponent is also known as the index or power.

identity (or identity element) An element in a set that, when combined with another element in a binary operation, results in the second element remaining the same. For example, in the set of positive integers where the operation is addition, the identity is 0. In the same set where the operation is multiplication, the identity is 1.

incompleteness theorem Theorem proposed by Kurt Gödel, in which he stated that any system of mathematical rules that includes the rules of arithmetic cannot be complete. This means that it will always be the case that there are mathematical statements that cannot be proved or disproved using just the rules of the system.

integer Any whole number, that is the counting numbers 1, 2, 3, 4, 5, and so on, 0, or the negative whole numbers.

intersection In set theory, the name for the set which contains only those elements common to two or more other sets. For example, given two sets A and B, the intersection describes the set of entities that belong precisely to both A and B.

inverse (or inverse operation) An operation that reverses the effect of another operation. For example, the inverse of addition is subtraction, and vice versa, while the inverse of multiplication is division, and vice versa.

operation Any formal set of rules that produces a new value for any input value or set of values. The four most common operations in arithmetic are addition, multiplication, subtraction, and division.

polynomial An expression using numbers and variables, which only allow the operations of addition, multiplication, and positive integer exponents, that is, x^2. (See also Polynomial Equations, page 80.)

property A characteristic or attribute that can be applied to an entity. Properties don't have to be physical in nature; for example, the numbers 2, 4, 6, and 8 share the property of being even numbers.

quintic polynomial Polynomial equation in which the highest exponent of an occurrence of a variable is 5.

real number Any number that expresses a quantity along a number line or continuum. Real numbers include all the rational numbers (that is, numbers that can be expressed as a ratio or fraction; including the positive and negative integers and decimals), the irrational numbers (those numbers that cannot be written as a vulgar fraction, such as $\sqrt{2}$), and the transcendental numbers (such as π).

term A single number or variable, or a combination of numbers and variables that are divided by an operation such as $+$ or $-$ to form an expression. For example in the equation $4x + y^2 - 34 = 9$, $4x$, y^2 and 34 are terms.

variable A quantity that can change its numerical value. Variables are often expressed as letters such as x or y, and are often used as placeholders in expressions and equations such as $3x = 6$, in which 3 is the coefficient, x is the variant, and 6 is the constant.

THE VARIABLE PLACEHOLDER

the 30-second math

3-SECOND SUM
In algebra, symbols such as x and y are used to represent unknown numbers, or quantities whose values can change.

3-MINUTE ADDITION
Within mathematics, algebra allows general laws of numbers to be expressed. For example, start with two numbers: 4 and 5. Then multiply each of them by a third number, 3, giving 12 and 15. Then add up the results: 27. This produces the same answer as adding together the two original numbers $(4 + 5 = 9)$ and then multiplying by the third $(9 \times 3 = 27)$. This is true for any three initial numbers. This law can be expressed algebraically: $(x + y)z = xz + yz$.

Scientists are always discussing numbers, but they often want to do so without pinning down their exact values. For example, we might want to say that in a certain room there are twice as many women as men. It is possible to express this relationship between the two numbers without knowing their values, by using a placeholding symbol such as x. If the (as yet unknown) number of men in the room is x, then the number of women is 2 times x (usually abbreviated to $2x$). If we later establish that $x = 7$, say, we can then substitute this value in order to get the number of women: $2x = 14$. This abstract, algebraic approach is useful throughout science. If a car travels at a constant speed s, over a distance d, for a time t, then a certain relationship must hold between the numbers s, d and t, whatever their specific values—namely, the speed must be equal to the distance divided by the time, that is $s = d/t$. This is a general law, but substituting in numerical values allows calculations in specific cases. If we subsequently discover any two of the values (such as $d = 10$ and $t = 2$) we can then use this formula to find the third $(s = {}^{10}/_2 = 5)$.

RELATED THEORIES
See also
THE EQUATION
page 78
POLYNOMIAL EQUATIONS
page 80

3-SECOND BIOGRAPHIES
DIOPHANTUS
c. 200–284

ABU 'ABDALLAH MUHAMMAD IBN MUSA AL-KHWARIZMI
c. 770–c. 850

ABU KAMIL SHUJA
c. 850–930

OMAR KHAYYAM
1048–1131

BHASKARA
1114–1185

30-SECOND TEXT
Richard Elwes

In algebra x marks the spot, when you've got an unknown number.

THE EQUATION

the 30-second math

The most important symbol
in mathematics is $=$. This asserts that the two quantities on either side of it are equal. An equation is any statement of this form. Of course, obvious equations like $7 = 7$ are rather uninteresting. But equations can be informative when the equality is less immediate. One famous example is $E = Mc^2$, the equation in physics which asserts that the energy (E) contained within an object is equal to its mass (M) multiplied by the speed of light (c) twice. Many fundamental laws in physics are in the form of equations. A common type of equation involves an unknown number. If x is a number such that $2x + 1 = 9$, that is to say "2 times x plus 1 equals 9," then this equation contains enough information to pin down x exactly. There is only one possible value of x if that equation is true. With any equation, the primary rule is "always do the same thing to both sides in order to keep it true." So if you want to subtract 1 from one side, you must do it to both: $2x = 8$. Similarly, when dividing one side by 2, you must do it to both: $x = 4$. This is now the "solution" to the original equation.

3-SECOND BIOGRAPHIES
EUCLID
c. 325–265 BCE

DIOPHANTUS
c. 200–284

ABU 'ABDALLAH MUHAMMAD IBN MUSA AL-KHWARIZMI
c. 770–c. 850

ABU BEKR IBN MUHAMMAD IBN AL-HUSAYN AL-KARAJI
c. 953–1029

ALBERT EINSTEIN
1879–1955

30-SECOND TEXT
Richard Elwes

3-SECOND SUM
Whenever two quantities are asserted to be equal, we have an equation. Most scientific statements take this form.

3-MINUTE ADDITION
Equations may do more than simply assert that numbers are equal to each other; they can deal with more sophisticated objects. "Differential equations" say that two different geometrical quantities are actually the same. Einstein's "field equation" in general relativity says that the way matter moves within a region of space is equal to the way that space itself is curved. Understanding the geometry of the universe involves solving this equation.

All things being equal, science is built from equations, from kindergarten arithmetic to relativity theory.

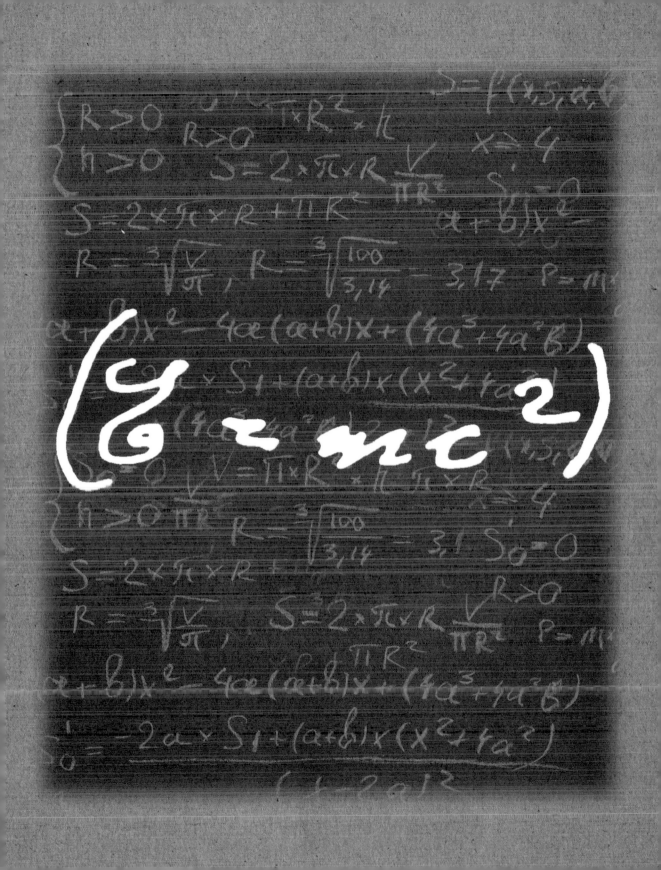

POLYNOMIAL EQUATIONS

the 30-second math

High school algebra students

learn to solve equations such as $3x^2 + 5x - 1 = 0$. This is an example of a polynomial equation, which by definition involves a sum of terms (for example, $3x^2$) in which a variable (for example, x) is raised to positive integer power (in this case, 2). The above equation is a second-degree equation, or quadratic, since the highest exponent (that is, the number of times the base number is to multiply itself) is 2. Thornier operations—involving fractional exponents, trigonometric and exponential functions—aren't allowed in a polynomial, which puts polynomials among the most basic of all equations. Methods for solving quadratic polynomials (finding values for the variable that render the equation consistent) were discovered in ancient times independently in several parts of the globe. The culmination of these efforts was the quadratic formula, which allows one easily to find the exact solutions. A complete solution of cubic (degree 3—where the highest exponent is 3) and quartic (degree 4) equations had to wait until 16th-century Italy, when mathematicians found formulas similar to the quadratic formula but more complicated. The search for a quintic (degree 5) formula ended more than 200 years later, when Niels Abel proved one of the first great negative results in mathematics: there is no general formula for solving a degree 5 or higher polynomial equation!

3-SECOND SUM
Polynomials are the formulas you get using numbers and variables, allowing only the operations of addition, multiplication and positive integer exponents (such as x^2).

3-MINUTE ADDITION
Being geometrically inclined, the ancient Greeks solved quadratic equations by intersecting lines and circles constructed with straightedge and compass. The geometry of shapes defined by polynomial equations in more than one variable, known as *algebraic geometry*, is a central area of current mathematical research. And in science, the paraboloid, given by the 3-variable polynomial equation $z = x^2 + y^2$, defines a shape useful for satellite dishes and car headlights.

RELATED THEORIES
See also
RATIONAL & IRRATIONAL NUMBERS
page 16
FUNCTIONS
page 46
THE VARIABLE PLACEHOLDER
page 76

3-SECOND BIOGRAPHIES
NICCOLÒ TARTAGLIA
1499/1500–1557

GIROLAMO CARDANO
1501–1576

NIELS ABEL
1802–1829

EVARISTE GALOIS
1811–1832

30-SECOND TEXT
Jamie Pommersheim

Polynomial equations create beautiful three-dimensional shapes.

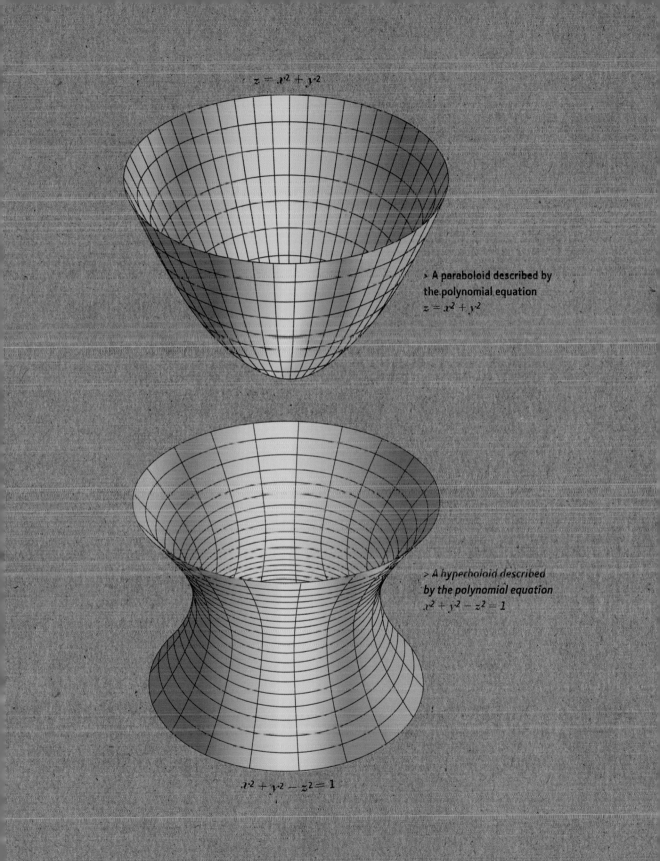

$$z = x^2 + y^2$$

> A paraboloid described by
the polynomial equation
$z = x^2 + y^2$

> A hyperboloid described
by the polynomial equation
$x^2 + y^2 - z^2 = 1$

$$x^2 + y^2 - z^2 = 1$$

c. 770–780
Born in Khwarizm,
modern Uzbekistan

825
Wrote *On the Calculation
with Hindu Numerals*

c. 830
Wrote *The Compendious
Book on Calculation
by Completion and
Balancing*

830
Produced a map of the
world

c. 850
Died

Mid 12th century
Robert of Chester
translated *The
Compendious Book
on Calculation by
Completion and
Balancing*

1126
Adelard of
Bath translated
al-Khwarizmi's
Astronomical Tables

12th century
Adelard of Bath
translated *On the
Calculation with Hindu
Numerals*

1857
*Algoritmi de numero
Indorum* (al-Khwarizmi
on the *Hindu Art of
Reckoning*) by Baldassarre
Boncompagni published

bedrock for mathematical study in the West. Little is known about his personal life; his family was Persian, and moved south to Baghdad (an Arab caliphate since the mid-seventh century) where he became a scholar in Caliph al Ma' mun's House of Wisdom (Bait al-Hikma), the library and academic institute at the heart of the Islamic Golden Age. Here al-Khwarizmi studied Greek and Sanskrit translations of scientific texts and works by Babylonian and Persian scholars. Although a formidable geographer, cartographer (he revised and corrected Ptolemy's *Geographica*, and wrangled 70 geographers to produce a map of the world for the caliph), and astronomer, his greatest and most invaluable contribution was to mathematics specifically, algebra, arithmetic, and trigonometry. He brought together techniques, methods, and concepts from India and farther east and added innovations and improvements of his own.

his 825 work *On the Calculation with Hindu Numerals*), Arabic numerals, tens and units, fractions, and the decimal point. He is probably best known as the father of algebra (although this again was a matter of synthesizing existing knowledge, then adding his own interpretation and techniques). In fact the word "algebra" comes from *al-jabr* (meaning "completion, restoring"), part of the title of his great work *The Compendious Book on Calculation by Completion and Balancing*, the first systematic solution of linear and quadratic equations. This had been commissioned by the caliph to be a practical, accessible work with real-world examples, offering solutions to problems in trade and commerce.

When al-Khwarizmi's work was translated into Latin in the 12th century, mathematics gained another new word. The Latinized form of his name is Algoritmi, from which is derived the word algorithm. There is also a crater named for him on the dark side of the moon.

ALGORITHMS

the 30-second math

The information revolution of the 20th century saw the rise of the computer. But computers are nothing without programs, and computer programs are nothing more than realizations of mathematical objects called algorithms. An algorithm is not complex; it is just a list of instructions for carrying out a task, where every step is completely unambiguous, and so can be carried out by an unthinking agent. The word algorithm derives from al-Khwarizmi, who discovered foolproof procedures for solving certain equations. Many mathematicians developed similar ideas over the centuries, but it was not until the work of Alan Turing and Alonzo Church in the 1930s that the notion of an algorithm was finally made precise. Turing considered a device comprising a paper tape, along which a "Turing machine" crawled, writing and erasing symbols according to strict internal rules. Turing used this theoretical contraption to demonstrate that no single procedure could ever answer every mathematical question. Even among the whole numbers there are some "uncomputable" problems. This echoed Gödel's incompleteness theorem, and was just as shocking to mathematics. But it was when the Turing machine crossed from the abstract mathematical domain into the real world that the digital computer was born.

RELATED THEORIES
See also
POLYNOMIAL EQUATIONS
page 80
HILBERT'S PROGRAM
page 142
GÖDEL'S INCOMPLETENESS
THEOREM
page 144
AL-KHWARIZMI
page 82

3-SECOND BIOGRAPHIES
ALONZO CHURCH
1903–1995

STEPHEN KLEENE
1909–1994

ALAN TURING
1912–1954

STEPHEN COOK
1939–

30-SECOND TEXT
Richard Elwes

3-SECOND SUM
Algorithms were conceived as theoretical procedures for carrying out mathematical tasks. They are now in constant use in computers around the world.

3-MINUTE ADDITION
The biggest questions in computer science concern how fast algorithms can run. For instance, start with two large prime numbers, and multiply them together. The challenge is to discover the two original numbers from the final result. There is an algorithm to do this, but it may take millions of years, even on the fastest modern processor. Is there a quicker way? No one knows. But we hope not, because this is what keeps our bank accounts safe online!

Every computer program encodes an algorithm, an idea dating back to the ninth century.

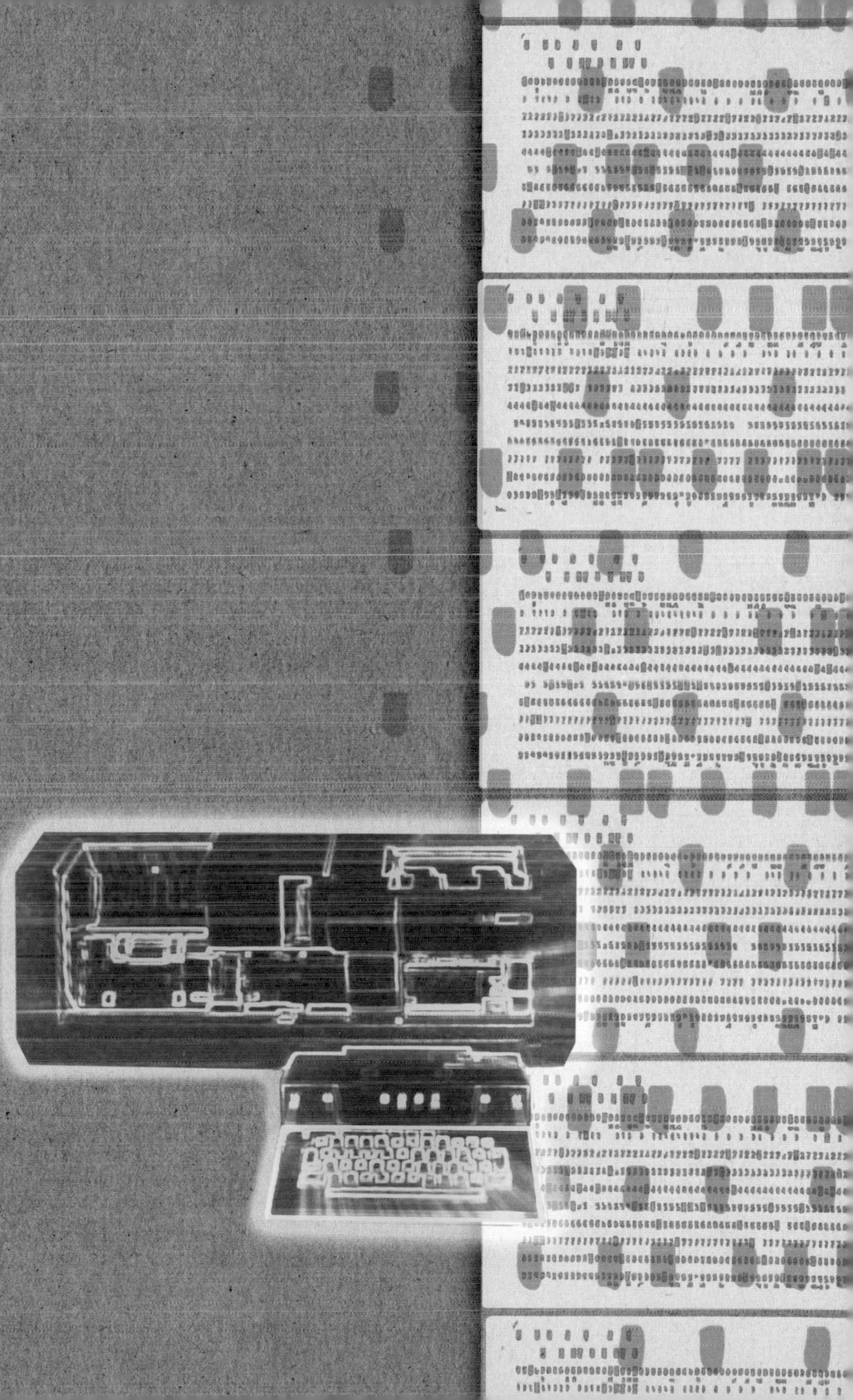

SETS & GROUPS
the 30-second math

Collecting and categorizing

objects is a key element of mathematics. Collections of objects (sets) allow us to define the common properties of the things we are studying. Creating unions of sets (combining them by taking one of each of their objects into a new set), or intersections (taking only what is common to both), helps us to refine their properties. As with numbers, we can combine objects in a set to make other objects in the same set. A group is a set with some special properties. (1) Any two objects in the set can be combined, via an operation (addition, for example), and the combination of any two objects must already be in the set. (2) There is a special object in the set called the identity, with the property that any object combined with the identity leaves the object unchanged—for example, 0 is the additive identity since you can add it to any other integer and the value will not change. And (3) to every group object there is another group object called its inverse. Any object combined with its inverse is the identity. Think of all of the integers with addition as the combining operation and 0 as the identity and you get the idea—for example, $5 + -5 = 0$.

RELATED THEORIES
See also
FUNCTIONS
page 46
RINGS & FIELDS
page 88

3-SECOND BIOGRAPHIES
JOSEPH-LOUIS LAGRANGE
1736–1813

NEILS HENRIK ABEL
1802–1829

ÉVARISTE GALOIS
1811–1832

ARTHUR CAYLEY
1821–1895

GEORG CANTOR
1845–1918

BENOÎT MANDELBROT
1924–2010

30-SECOND TEXT
David Perry

3-SECOND SUM
Any collection of objects is a mathematical set. A group is created by combining objects in a set to make other objects in the set.

3-MINUTE ADDITION
Although we've been thinking of numbers as our objects, things can become more interesting when you introduce different types of elements as your objects. Indeed, the famous Circle of Fifths in music theory is the set of the 12 major scales. It can be given a group structure called a cyclic group.

Venn diagrams provide visual aids to understanding the relationships between several sets.

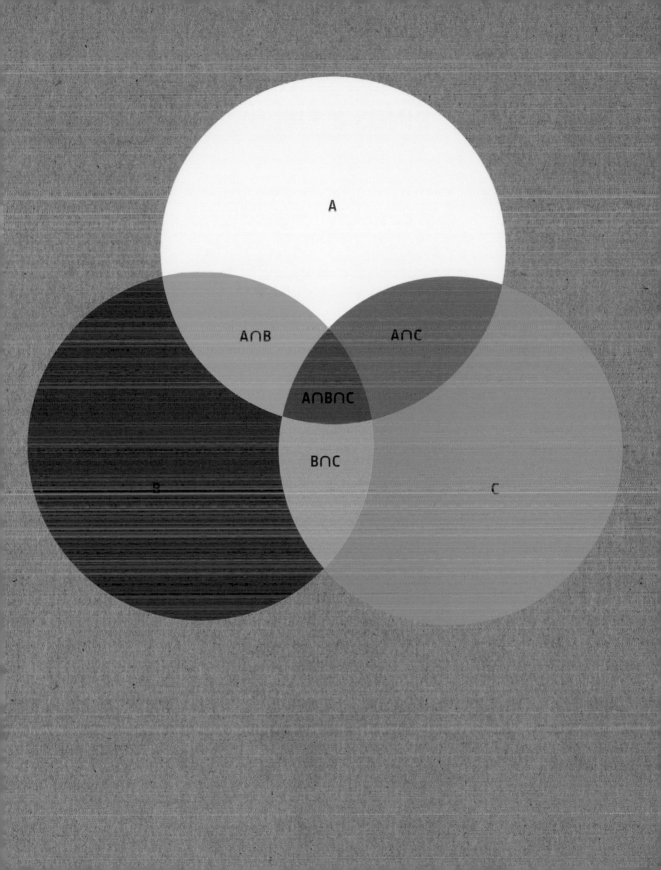

RINGS & FIELDS
the 30-second math

Arithmetic with integers involves two fundamental operations: addition and multiplication (after which one learns about subtraction and division as well). In school, we learn that the sum $1 + 4 + 9 + 16$ requires no parentheses because we can start anywhere in this sum, even rearranging the terms, and always get the same answer (because addition is associative and commutative). We learn how the operations interact when we learn the distributive property of integers: $a \times (b + c) = a \times b + a \times c$. Many sets possess these same useful properties exhibited by integers. We won't list them here, but we give all sets with these properties a name: rings. The set of real numbers is also a ring, although it has an additional useful property integers don't possess. With integers, although you can add or multiply two integers and obtain an integer, and you can also subtract two integers and get an integer, you cannot necessarily divide two integers and get an integer. On the other hand, you can divide any real number by any other real number (other than zero!) and get a real number. This distinction gives the set of real numbers the designation of field.

RELATED THEORIES
See also
ADDITION & SUBTRACTION
page 40
MULTIPLICATION & DIVISION
page 42
POLYNOMIAL EQUATIONS
page 80
SETS & GROUPS
page 86
SQUARING THE CIRCLE
page 104

3-SECOND BIOGRAPHIES
ÉVARISTE GALOIS
1811–1832

RICHARD DEDEKIND
1831–1916

EMMY NOETHER
1882–1935

30-SECOND TEXT
David Perry

3-SECOND SUM
The set of integers has nice properties that earn it the designation of ring. The set of real numbers is even nicer, and is called a field.

3-MINUTE ADDITION
Rings and fields were historically important because they allowed mathematicians to translate some classical problems into a brand-new language. This new language allowed for long-desired proofs that the circle cannot be squared, nor the cube doubled, nor an arbitrary angle trisected using only straightedge and compass. It also allowed mathematicians to prove that in spite of the existence of a quadratic formula—and cubic and quartic formulas—no such formula could exist for quintic polynomials.

The distributive property concerns how addition and multiplication interact —sets with these properties are called rings.

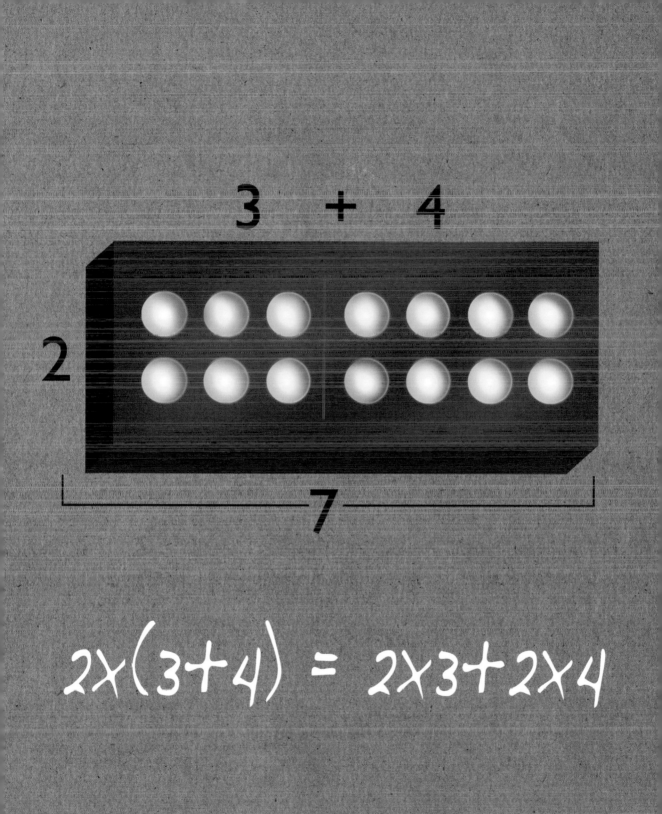

GEOMETRY & SHAPES

axiom A proposition or statement that is self-evidently true or has been accepted as true without proof.

circumference The boundary line or perimeter of a curved figure, usually used in reference to a circle.

conic section A curved figure created by the intersection of a plane with a circular cone. A conic section can either be a circle, an ellipse, a parabola, or a hyperbola, depending on the angle at which the plane intersects the cone.

constant A number, letter, or symbol on its own that represents a fixed value. For example, in the equation $3x - 8 = 4$, 3 is the coefficient, and x is the variable, while 8 and 4 are the constants. However, the term is usually more closely associated with symbols such as π or e.

crank In mathematical circles, the term "crank" is affectionately applied to those people who refuse to accept proven mathematical theorems.

diameter A straight line passing through the centre of a circle or sphere, running from one side to the opposite side. More generally, the largest distance between any two points in the same figure.

dodecahedron Term usually used to describe a regular polyhedron with 12 faces, each of which forms a pentagon. Dodecahedrons are one of the five Platonic solids. A rhomboid dodecahedron is an example of an irregular dodecahedron.

Euclidean geometry The study of lines, points, and angles in planes and solids. Named after the ancient Greek mathematician Euclid of Alexandria, Euclidean geometry is the entire mathematical system of rules and laws based around five axioms that he postulated in his work *The Elements*.

Galois theory Methods by which algebraic structures, known as groups, can be used to solve algebraic equations.

geometry The branch of mathematics that deals primarily with shapes, lines, points, surfaces, and solids.

hexagon Polygon with six straight sides and six angles.

hyperbolic geometry A form of non-Euclidean geometry in which the parallel postulate in Euclidean geometry is replaced with the postulate that there are at least two lines in the plane that do not intersect a given line. In hyperbolic geometry, the sum of the angles of a triangle is less than 180º. See *Euclidean geometry*.

hypotenuse In a right-angled triangle, the side opposite the right angle. The hypotenuse plays a fundamental role in the Pythagorean theorem. See *Pythagorean theorem*.

icosahedron A regular polyhedron made up of 20 faces, each of which forms an equilateral triangle. Icosahedrons are one of the five Platonic solids.

lemma A mathematical truth that is used to support a more important mathematical truth, such as a theorem. Also, a stepping stone to a larger mathematical truth.

number theory The branch of mathematics that deals primarily with the properties and relationships of numbers, with particular attention being given to positive integers.

pentagon Polygon with five straight sides and five angles.

pentagram Five-pointed star made up of five straight lines.

polyhedron Any solid with four or more faces made up of polygons. In regular polyhedrons, such as the five Platonic solids, the faces are made up of regular polygons.

proposition A statement of a theorem or problem. Propositions are usually accompanied by a demonstration of their truth (a proof).

Pythagorean theorem A theorem attributed to Pythagoras which states that for a right-angled triangle, the square of the length of the hypotenuse (the side opposite the right angle) is equal to the sum of the squares of the lengths of the other two sides. It is commonly formulated as $a^2 + b^2 = c^2$.

radius The distance from the centre of a circle to its edge. The radius is half the value of the diameter.

theorem A mathematical fact or truth, which has been established as a logical consequence or arising from previously accepted mathematical facts or axioms.

transcendental number Any number that cannot be expressed as a root of a non-zero polynomial with integer coefficients; in other words, non-algebraic numbers. π is the best-known transcendental number, and following the opening definition π therefore could not satisfy the equation $\pi^? = 10$. Most real numbers are transcendental.

EUCLID'S ELEMENTS

the 30-second math

Euclid was a Greek mathematician
who lived and taught in the Egyptian city of
Alexandria around 300 BCE. He is revered not
only for his specific theorems concerning
triangles, circles and prime numbers, but also
for his entire approach to mathematical
thought in providing definitions, identifying
the postulates being assumed, then carrying
forward the logical consequences of those
basic assumptions, lemma by lemma, theorem
by theorem. He provided a methodology for
mathematical reasoning that served as an
inspiration for the next 22 centuries of geometry
instruction throughout the world. Although
much of his most celebrated, 13-book work, *The
Elements*, concerns geometry (in Book I, Euclid
proves the Pythagorean theorem, while he
explains the construction of the five Platonic
solids in Book XIII), Euclid made a three-book
excursion into number theory. In Book VII he
explains how to find the greatest common
divisor of two integers, detailing an algorithm
that bears his name. In Book IX he returns to
the Pythagorean theorem and provides a
formula that generates whole numbers whose
squares add up to the square of another whole
number, such as $3^2 + 4^2 = 5^2$, giving lengths
of sides of a right-angled triangle.

RELATED THEORIES
See also
PRIME NUMBERS
page 22
SQUARING THE CIRCLE
page 104
PARALLEL LINES
page 106
PLATONIC SOLIDS
page 114

3-SECOND BIOGRAPHIES
PYTHAGORAS
C. 570–C. 490 BCE

EUCLID
fl. 300 BCE

30-SECOND TEXT
David Perry

*A proof of the
Pythagorean triple.
Congruent triangles
can be used to show
that the grey square
has the same area as
the yellow rectangle
and that the red square
has the same area as
the blue rectangle.*

Ἐν ἄρα τοῖς ὀρθογωνίοις τριγώνοις τὸ ἀπὸ τῆς τὴν ὀρθὴν γωνίαν ὑποτεινούσης πλευρᾶς τετράγωνον ἴσον ἐστὶ τοῖς ἀπὸ τῶν τὴν ὀρθὴν [γωνίαν] περιεχουσῶν πλευρῶν τετραγώνοις· ὅπερ ἔδει δεῖξαι.

PI—THE CIRCLE CONSTANT

the 30-second math

3-SECOND SUM

'*Quantitas, in quam cum multiplicetur diameter, proveniet circumferentia*' —"The quantity which, when the diameter is multiplied by it, yields the circumference." That's π (or pi) to you and me.

3-MINUTE ADDITION

In piphilology, a "piem" is a poem devised so that the letter-length of each word coincides with the decimal expansion of π. Sir James Jeans started the game: "How I want a drink, alcoholic of course, after the heavy lectures involving quantum mechanics." Get it? The *Cadaeic Cadenza*, a short story written by Mike Keith in 1996, is said to be written in pilish. It is a piem in prose, whose word-length is 3,835!

Arguably the best- and longest-known easy-to-see-but-hard-to-calculate mathematical constant, the irrational (transcendental) number $\pi = 3.1415926535897...$ was known to all of the ancient civilizations due to its simple relationship to the circle. It is the ratio of the circumference of a circle to its diameter. It is widely thought that the choice of the Greek letter for the constant came from the word for "perimeter" (περίμετρος), and it is sometimes called Archimedes' constant due to his famous attempts to calculate it. Indeed, from the circle approximations by inscribed or circumscribed polygons of people like Archimedes or the Chinese mathematician Liu Hui, through the finite sums of an infinite number of fractions via the calculus of Leibniz, to fascinating equations like the formulas of the Indian mathematician Ramanujan, π has most likely spawned more mathematical study than any other single concept and plays a central role in almost every natural and social science. Always the enigmatic number, π has spawned contests for humans to recall its decimal digits in order and for computers to calculate ever more accurate approximations. Celebrations of the number include π-day (March 14, or 3/14), a now global phenomenon, and the development of a new (serious but rather humorous) field of study called π-philology, or piphilology.

RELATED THEORIES

See also
RATIONAL & IRRATIONAL NUMBERS
page 16
TRIGONOMETRY
page 102
SQUARING THE CIRCLE
page 104

3-SECOND BIOGRAPHIES
PYTHAGORAS
C. 570–C. 490 BCE

ARCHIMEDES
C. 287–212 BCE

ISAAC NEWTON
1643–1727

WILLIAM JONES
1675–1749

30-SECOND TEXT
Richard Brown

Archimedes' method of drawing a series of polygons inside and outside a circle enabled him to work out the approximate value of π.

THE GOLDEN RATIO

the 30-second math

3-SECOND SUM
The number for which the ratio of the sum of two parts to the larger part is the same as the ratio of the larger part to the smaller part.

3-MINUTE ADDITION
The golden ratio is often cited as playing an aesthetic role in art, architecture and design dating back to the pyramids of the ancient Egyptians, the temples of classical Greece, through to the paintings of Leonardo da Vinci, and even to today's iPod. However, although there are examples of artists and designers deliberately incorporating the ratio into their work (the architect le Corbusier, for example), there are many that question the golden ratio's artistic significance.

If you divide a line into larger and smaller parts a and b so that the sum of the two parts divided by the larger part is equal to the larger part divided by the smaller part, that is, $(a + b)/a = a/b$, you obtain the golden ratio. It is also known as the golden section, golden mean and divine proportion and is denoted by the Greek letter phi (ϕ), which is the irrational number yielded through solving the equation: $\phi = (1 + \sqrt{5})/2 = 1.6180339887498...$ For mathematicians, it is interesting to note that ϕ also satisfies $\phi^2 = 1 + \phi$ and $1/\phi = \phi - 1$. The golden ratio is also the measure of the diagonal of a regular pentagon with sides of length 1. The pentagram, a figure formed by the diagonals of a pentagon, had mystical associations for Pythagoras and his followers. Artists and architects use the golden ratio to create proportions that are pleasing to the eye. The Fibonacci sequence, 1, 1, 2, 3, 5, 8, 13, 21, 34, ... has the property that the ratio of two consecutive numbers approaches ϕ as the numbers become large. The golden rectangle, with sides in proportion to the golden ratio, is found in both the dodecahedron and icosahedron. A golden spiral is formed by fitting quarter-circular arcs in squares with edge lengths that diminish sequentially by ϕ.

30-SECOND TEXT
Robert Fathauer

A series of squares with relative side lengths scaled by the golden ratio fit neatly together in a spiralling configuration. Quarter-circular arcs inscribed in the square form the golden spiral.

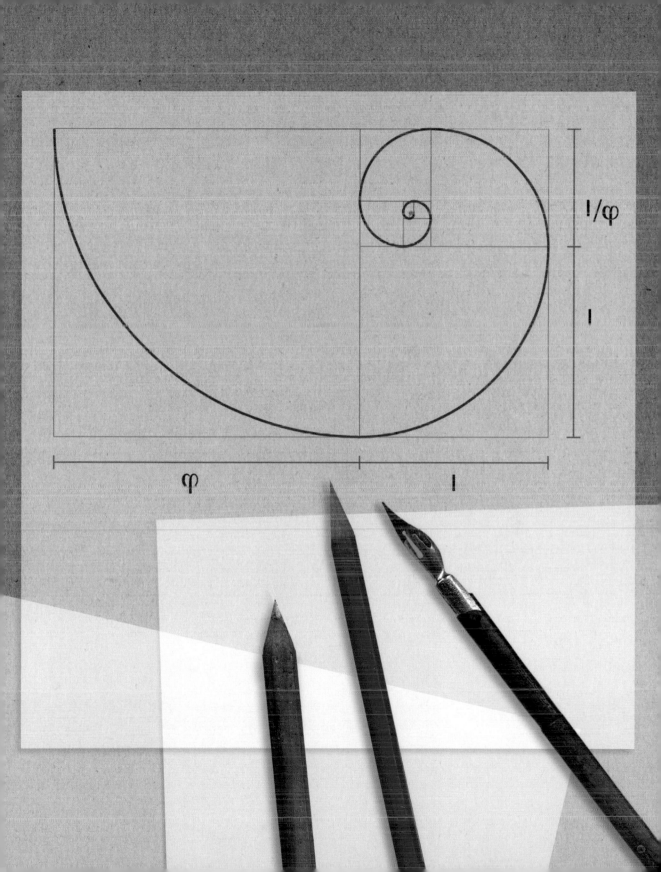

C. 570 BCE
Born in Samos

C. 530 BCE
Moved to Croton, southern Italy

C. 490 BCE
Died, probably in Metapontum

C. 200–250 CE
Diogenes Laertius, author of *Lives and Opinions of Eminent Philosophers*

C. 234–305 CE
Porphyry, author of a *Life of Pythagoras*

C. 245–325 CE
Iamblichus, author of *On the Pythagorean Life*

PYTHAGORAS

Most non-mathematicians have

a schoolroom memory of Pythagoras' theorem, and that is what he is best remembered for in the modern mind. The man himself was far more enigmatic and a whole academic industry has grown up around what is known as the "Pythagorean Question," which tries to disentangle the real, historical Pythagoras and his achievements from layers of myth, spin, and almost hagiographic legend that have accreted around him. He never wrote anything down, and nor did his contemporaries, so hardly anything is known about him and to his many followers he became a semi-divine, mystical character—a King Arthur of the ancient world.

A mysterious and charismatic figure, he is said to have had a golden thigh, to have performed wonders and possessed the shamanic ability to be in two places at once. Pythagoras believed that the soul was immortal and went through several reincarnations, and he was the founder of an esoteric religious cult, much admired for its principled and rigorous austerity, and significant enough to be

persecuted by the political establishment. We know this much because devoted followers—Pythagoreans were a flourishing sect until the fifth century CE—began to write about him some 150 years after his death. They rewrote history and glorified his achievements, maintaining that Pythagoras was the source of all Aristotelian and Platonic ideas. The many treatises put out under his name are forgeries. However, as far as math is concerned, although Pythagoras recognized a divine and mystical meaning to numbers and their relationship with each other, it is unlikely that he ever proved his theorem. The only piece of evidence that he studied geometry is based on retrospective propaganda. We now know that the theorem was known to Babylonian scholars in arithmetical form, although they did not prove it either, so it could be that the seed of the story was that Pythagoras was simply recognized as passing on a significant and elegant piece of mathematical knowledge.

TRIGONOMETRY

the 30-second math

A right-angled triangle has the property that the angles are related to ratios of the side lengths. This relationship forms the basic "sine function" (and its cousins such as "cosine"), where the sine of an angle equals the ratio of the opposite side length to that of the hypotenuse (the side opposite the right angle). Knowing how to calculate length from angle measurements had enormous practical implications to ancient astronomers and explorers, from the Sumerians and the ancient Greeks to the Indians and Persians. Hipparchus, the second century BCE Greek astronomer, is considered the "father of trigonometry." Modern scientists view "trig" functions more broadly. Points on a circle can be pinpointed via a right-angled triangle; if the radius is 1, the coordinates of a point on the circle are the cosine and sine of the angle Θ. As Θ is increased, the y value (sine of Θ) first increases then decreases, becomes negative and returns to zero. As Θ continues to increase beyond 2 it repeats this cycle over and over, so that a graph of the sine of Θ vs. Θ has a periodic (repeating) wave shape. Hence all phenomena that look or act wavelike, from radiation in physics, through sound in music, to oceanography, medical imaging, and much of engineering and architecture, can be studied using the basic trig functions such as sine and cosine.

RELATED THEORIES
See also
FUNCTIONS
page 46
CALCULUS
page 50
PI – THE CIRCLE CONSTANT
page 96
GRAPHS
page 108

3-SECOND BIOGRAPHIES
HIPPARCHUS
C. 190–120 BCE

PTOLEMY
C. 90–165 CE

LEONHARD EULER
1707–1783

30-SECOND TEXT
Robert Fathauer

3-SECOND SUM
Trigonometry is the study of the relationships between the angles of a triangle and the lengths of its sides. It is fundamental to all modern science.

3-MINUTE ADDITION
In plane trigonometry, widely taught in school, all triangles have angles that add up to 180°. Spherical trigonometry, however, is what is used for astronomy and was of greater interest to ancient civilizations. On a sphere, the angles of a triangle add up to more than 180°. In fact, with one point at the North Pole and two other points on the Equator (one a quarter turn away from the other), all three angles of the resulting triangle are 90°!

The cosine and sine functions are defined as the x and y coordinates of the point at which a line at angle Θ from the x-axis intersects the unit circle.

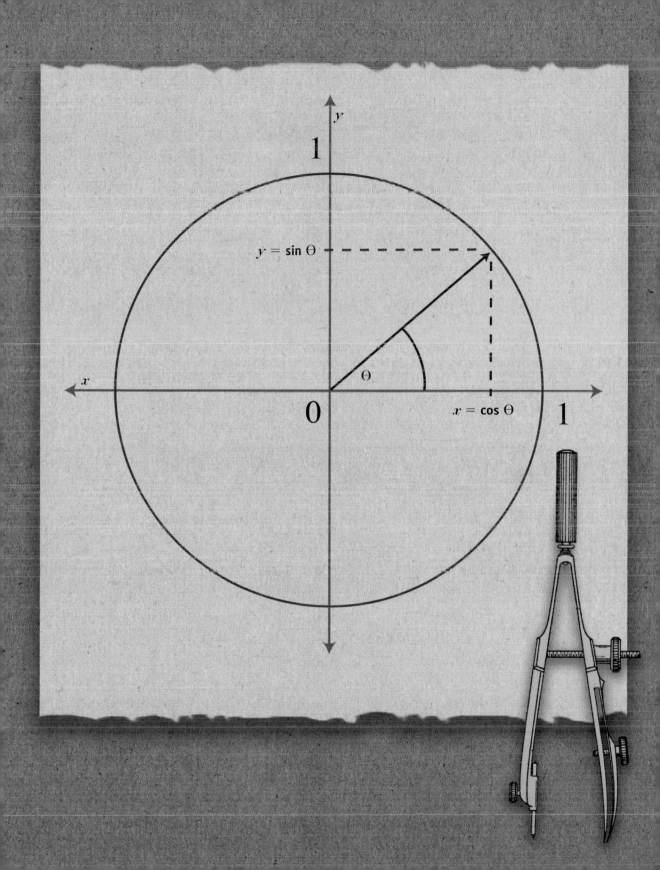

SQUARING THE CIRCLE

the 30-second math

The ancient Greeks thought of all numbers as lengths, so their math was almost exclusively done geometrically. Dividing a number by two was seen as a geometric construction. First, consider the number as the length of a line segment. Then, use the tools of geometry, namely a straightedge and a compass, to divide that segment in half. You have achieved division by two. Starting with a circle, one can attempt to construct a square whose area is the same as that of the circle. Thousands of years ago mathematicians came close to "squaring the circle," but the early attempts relied on the assumption that π can be expressed as the ratio of two whole numbers. Not only is π now known to be irrational, it was proven to be transcendental in the 19th century. Centuries earlier, mathematicians had separately shown that transcendental numbers could not be constructed with straightedge and compass, resolving the issue definitively. Attempts at solutions had wonderful unforeseen benefits, however. Conic sections were invented by Menaechmus to solve these problems, as were abstract algebra and Galois theory, subjects of immense importance to mathematics today.

3-SECOND SUM
With basic tools, the task of drawing a square with the same area as a given circle seems quite straightforward. Alas, mathematicians know that this task is impossible.

3-MINUTE ADDITION
The tradition of performing geometric constructions allowing only the use of straightedge and compass is founded in the axioms codified in Euclid's *Elements*. The limitations of what one can do with these tools are built into the tools themselves. This does not deter a throng of amateur and professional mathematicians from claiming solutions to these impossible problems every year. Such men and women are affectionately called 'cranks' in the business. It seems to be in human nature to engage in quixotic quests.

RELATED THEORIES
See also
RATIONAL & IRRATIONAL NUMBERS
page 16
EUCLID'S ELEMENTS
page 94
PI—THE CIRCLE CONSTANT
page 96

3-SECOND BIOGRAPHIES
HIPPIAS OF ELIS
C. 450 BCE– ?

EUCLID
fl. 300 BCE

ARCHIMEDES
C. 287–C. 212 BCE

30-SECOND TEXT
David Perry

With only straightedge and compass, you can easily bisect an angle or construct a regular hexagon. You cannot, however, square the circle.

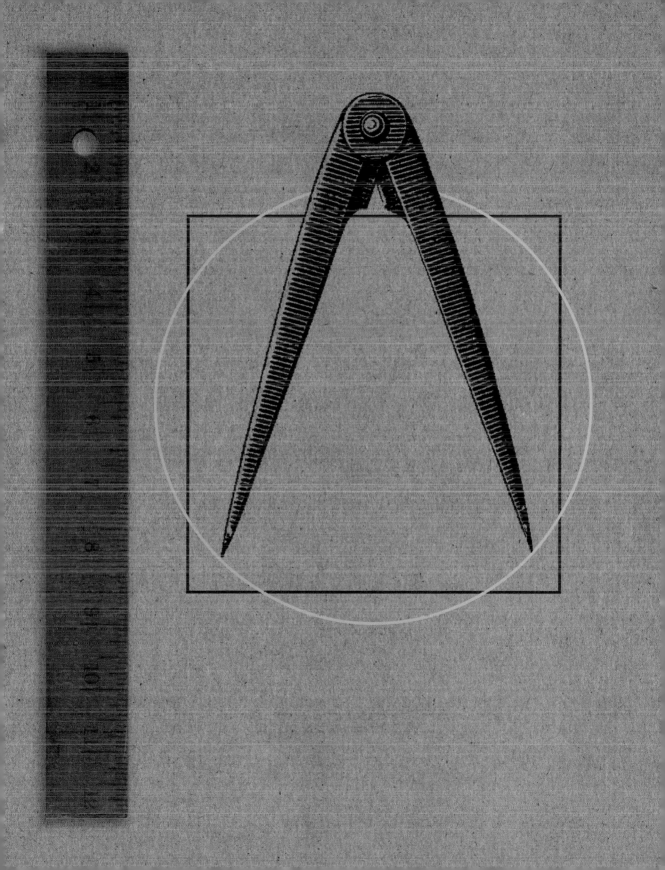

PARALLEL LINES

the 30-second math

Parallel lines took centre stage in Euclid's *Elements*, as he set about building the geometry of a two-dimensional plane from first principles. Euclid began with five fundamental laws of geometry. From these, he deduced facts familiar to generations of students, such as the corresponding angles theorem: if a pair of parallel lines is crossed by a third line, the angles in corresponding positions are equal. Euclid's fifth law, known as the "parallel postulate," says that if you draw a straight line, and then pick a point away from it, there is only one possible parallel that can be drawn through that point. Anyone who tries it on a piece of paper will easily be persuaded that it is true, but for thousands of years geometers tried to understand why it should be so. Many were convinced that it was a consequence of the other four, simpler, laws. It was not until the 19th century that Gauss, Bolyai, and Lobachevsky independently discovered an entirely new form of geometry satisfying the first four of Euclid's axioms, in which the parallel postulate failed. In this non-Euclidean "hyperbolic" geometry, there are infinitely many lines that could pass through a single point, parallel to a given line.

RELATED THEORY
See also
EUCLID'S ELEMENTS
page 94

3-SECOND BIOGRAPHIES
EUCLID
fl. 300 BCE

CARL-FRIEDRICH GAUSS
1777–1855

NICOLAI LOBACHEVSKY
1796–1856

JÁNOS BOLYAI
1802–1860

HERMANN MINKOWSKI
1864–1909

30-SECOND TEXT
Richard Elwes

3-SECOND SUM
Parallel lines are lines in the plane that continue for ever without meeting, like train tracks. The laws of parallel lines play a defining role in different forms of geometry.

3-MINUTE ADDITION
Hyperbolic geometry, with its abundance of parallel lines, fascinated geometers. It found a home in 20th-century physics in Einstein's new theory of special relativity. Hermann Minkowski showed that the geometry of the universe is fundamentally hyperbolic. It does not immediately appear that way, but when approached from the perspective that all speeds below the speed of light are equivalent, the hyperbolic nature of motion was revealed.

Parallel lines are among the most familiar patterns, and the keys to the most unfamiliar geometrical worlds.

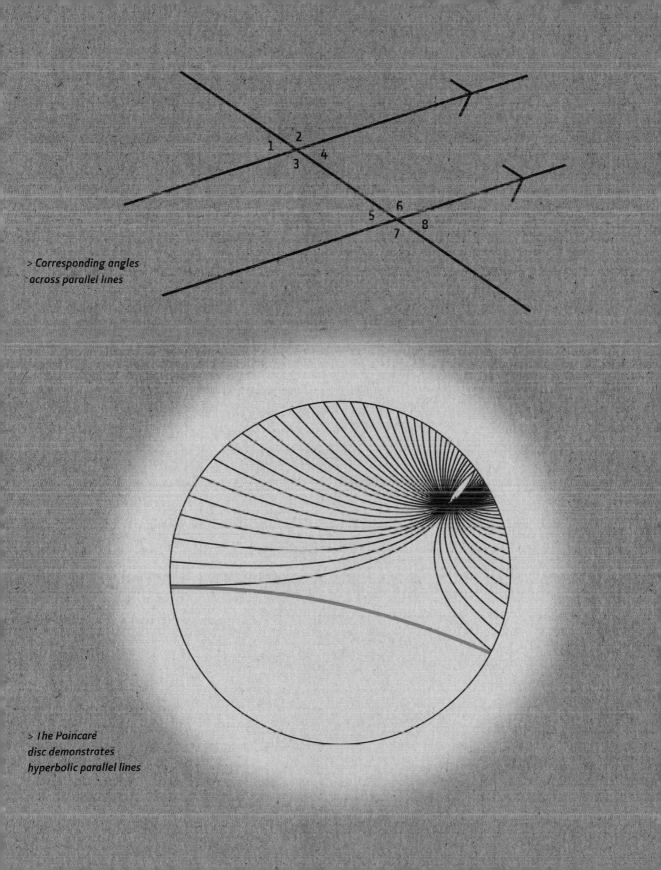

> *Corresponding angles across parallel lines*

> *The Poincaré disc demonstrates hyperbolic parallel lines*

GRAPHS

the 30-second math

3-SECOND SUM
A graph is a pictorial representation of the relationship between two or more variables.

3-MINUTE ADDITION
There are other coordinate systems in addition to the Cartesian one, such as polar coordinates, in which a radial coordinate r and angular coordinate Θ are specified. This allows a more ready solution of problems that deal with phenomena radiating from a point, such as antenna strength. More broadly, any map can also be considered a type of graph, because it relates data such as city and road names, elevation, and so on to geographical location.

In mathematics, graphs are most commonly used to depict mathematical functions. In other fields, from biology to business, graphs are primarily used to display data. Mathematical graphs are traditionally displayed on a set of two perpendicular axes labelled x and y in two dimensions. Any point in the plane can be specified via an "ordered pair" (x, y) specifying its distance from the y- and x-axes. The same concept is used for displaying information in three dimensions by adding a third axis conventionally labelled z. This system is known as Cartesian coordinates, after its discoverer, French mathematician and philosopher René Descartes. His contemporary, Pierre de Fermat, developed similar ideas independently. However, the invention of the graph may more properly be credited to Nicole d'Oresme, who three centuries earlier used horizontal and vertical axes to prove graphically a rule relating the distance covered by two objects moving at different rates. Descartes' realization of the potential of the graph was a seminal development in the history of mathematics, joining numbers and geometric figures. This made possible the representation of such figures with equations, bringing together algebra and geometry to create the field of analytical geometry.

RELATED THEORIES
See also
IMAGINARY NUMBERS
page 18
FUNCTIONS
page 46
CALCULUS
page 50
PARALLEL LINES
page 106

3-SECOND BIOGRAPHIES
NICOLE D'ORESME
c. 1320–1382

RENÉ DESCARTES
1596–1650

PIERRE DE FERMAT
1601–1665

30-SECOND TEXT
Robert Fathauer

The algebraic description of a particular ellipse (top) and the associated geometric figure graphed using Cartesian coordinates.

$$\frac{(x-1)^2}{4^2} + \frac{(y-2)^2}{3^2} = 1$$

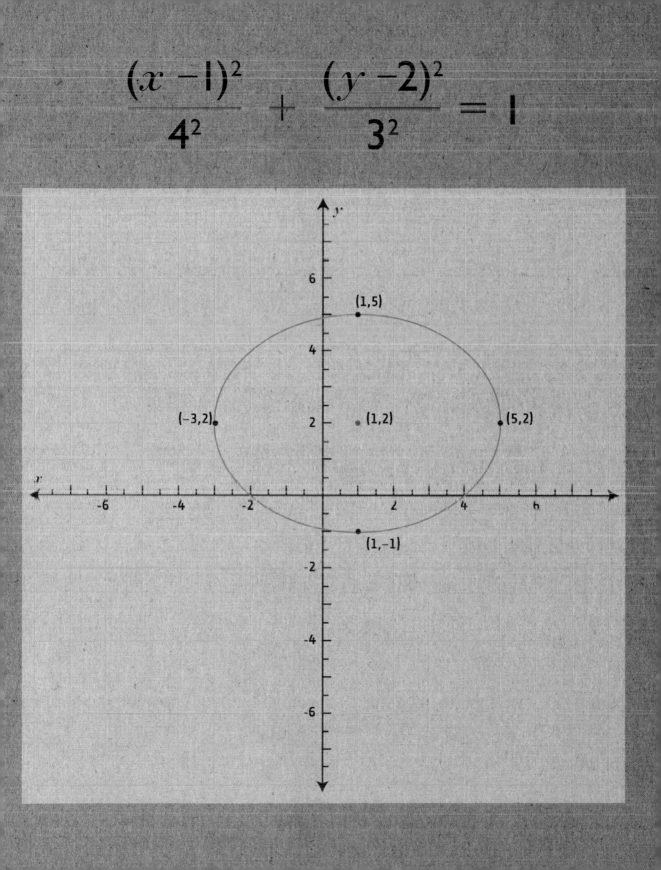

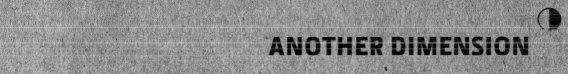

ANOTHER DIMENSION

axiom A proposition or statement that is self-evidently true or has been accepted as true without proof.

complex number Any number that comprises both real and imaginary number components, such as $a + bi$, in which a and b represent any real number and i represents $\sqrt{-1}$.

cube A solid with six sides, each of which is a regular square. Cubes are one of the five Platonic solids.

dodecahedron Term usually used to describe a regular polyhedron with 12 faces, each of which forms a pentagon. Dodecahedrons are one of the five Platonic solids. A rhomboid dodecahedron is an example of an irregular dodecahedron.

Euler characteristic In topology, term used to describe a shape's specific topological data. For three-dimensional polyhedra, it is based around the equation $V - E + F =$ Euler characteristic, in which V is the number of dots or vertices, E is the number of edges, and F is the number of faces.

factorial The product of a series of descending positive integers, such $6 \times 5 \times 4 \times 3 \times 2 \times 1$. The symbol for factorial is !, therefore $4! = 4 \times 3 \times 2 \times 1 = 24$.

fractional dimension The size or dimension of a fractal set may be a number between two natural numbers. The fractional dimension is a measure of the apparent self-similarity of a fractal.

icosahedron A regular polyhedron made up of 20 faces, each of which forms an equilateral triangle. Icosahedrons are one of the five Platonic solids.

iteration In fractal geometry, a repeated operation that performs the same task each time.

Jones polynomial In knot theory, a polynomial that describes certain characteristics of specific knots.

Klein bottle An object with an enclosed surface that has only one side and no edges. A Klein bottle cannot be visualized in three dimensions without self-intersections. It was named after the German mathematician Felix Klein, who first described the surface in 1882.

Koch snowflake In fractal geometry, one of the earliest fractals. Each side of an equilateral triangle undergoes an iteration (repeated operation) in which the middle third section of each side is replaced by a motif made up of two lines that form a point away from the main body of the triangle. The process is repeated infinitely.

octahedron Term usually used to describe a regular polyhedron made up of eight sides, each of which is an equilateral triangle. Octahedrons are one of the five Platonic solids.

polygon Any two-dimensional shape that has three or more straight sides.

polyhedron Any solid with four or more faces made up of polygons. In regular polyhedrons, such as the five Platonic solids, the faces are made up of regular polygons.

polynomial An expression using numbers and variables, that only allows the operations of addition, multiplication, and positive integer exponents, that is, x^2. (See also Polynomial Equations, page 80.)

tetrahedron A term that is usually used to describe a regular polyhedron made up of four sides, each of which is an equilateral triangle (hence its alternative name of a triangular pyramid). Tetrahedrons are one of the five Platonic solids.

torus In geometry, a doughnut-shaped figure.

vertex Any angular point or corner on a polygon or polyhedron.

PLATONIC SOLIDS

the 30-second math

Attaching different regular polygons together to form a solid is not so difficult. Think of the standard soccer ball, with its interlocking hexagons and pentagons. Doing so with only one polygonal shape, however, is more difficult. In fact, there are only five ways of doing this: the cube, with its six squares as sides; the tetrahedron, octahedron, and icosahedron, using four, eight, and 20 equilateral triangles respectively; and the dodecahedron, with its 12 pentagons. The ancient Greeks studied the collection extensively. Plato wrote about them in his dialogue *Timaeus*, and it is thought that Theatetus (Plato's contemporary) was the first to give a proof that there are no others. The idea? If more than two equilateral polygons meet, they must meet at a corner or vertex. At a corner, the sum of the angles of the polygons meeting there must add up to less than 360° (they cannot add up to more, and at 360° the shape would be flat). This is very restrictive. Any regular polygon with six or more sides has an angle of more than 120°. Three of those together wouldn't work! And there are precious few ways to have the remaining equilateral polygons meet like this. In fact, five is the precious few!

Meet the five Platonic solids—clockwise, from left: the cube, the tetrahedron, the dodecahedron, the icosahedron, and the octahedron.

TOPOLOGY
the 30-second math

3-SECOND SUM
Like geometry, topology, or rubber-sheet geometry, is the study of shapes. The difference is that topologists class two shapes as being the same if one can morph into the other.

3-MINUTE ADDITION
An important piece of topological data is a shape's "Euler characteristic." This involves drawing dots and connecting them with edges. On a sphere, we might draw two dots and two edges, dividing the surface into two faces. A fundamental fact states that with V dots, E edges and F faces, it must be true that $V - E + F = 2$ on any topological sphere. (A cube has $V = 8$, $E = 12$, and $F = 6$.) Meanwhile, a torus has Euler characteristic 0, meaning that $V - E + F = 0$.

In topology, a cube, a pyramid, and a sphere are all the same thing. The reason is that topologists are not interested in the fine geometrical details of a shape (length, area, angle, or curvature). Rather, topology focuses on the global aspects of a shape, and on information that overrides stretching and twisting (though never cutting or gluing). What features of a shape can survive this process? Typical topological information is the number and type of holes within a shape. For example, a lower case "i" consists of two parts separated by a gap, and topological morphing does not allow the gap to be closed. So, while "i" is equivalent to "j" and to the number "11," it is not equivalent to "L" or "3." Meanwhile, the hole in an "O" also cannot be removed, making it topologically identical to an "A" and a "9," but not to an "8" with its two holes. The London Tube Map is an example of topology in action. The precise geography of the city is eliminated, allowing the essential topological features such as the order of the stations and the intersection points of different lines to be displayed clearly.

RELATED THEORIES
See also
THE MÖBIUS STRIP
page 120
KNOT THEORY
page 130
POINCARÉ'S CONJECTURE
page 146

3-SECOND BIOGRAPHIES
LEONHARD EULER
1707–1783

JULES HENRI POINCARÉ
1854–1912

FELIX HAUSDORFF
1868–1942

MAURICE RENÉ FRÉCHET
1878–1973

LUITZEN EGBERTUS
JAN BROUWER
1881–1966

30-SECOND TEXT
Richard Elwes

What's the difference between a sphere and a cube? To a topologist, nothing.

EULER BRICKS

the 30-second math

It's easy to draw a rectangle in which the height and width are both whole numbers. But it's harder if we also want the diagonal distance across to be a whole number. If we try a square 1in wide by 1in high, then the diagonal comes out at around 1.41in – in fact $\sqrt{2}$in, by Pythagoras' theorem. The same thing happens with every square: if the sides are whole numbers, the diagonal can't be. This is also true for many rectangles, but there are some that work. One 3in wide and 4in tall has a diagonal of exactly 5in. Another has sides of 5in and 12in with diagonal 13in. Euler wanted a brick in which all the edges were whole numbers, as were the diagonals of each face. The first was discovered by Paul Halcke in 1719. It's 44 units high, 117 wide and 240 long, and its faces have diagonals of 125, 244 and 267. Since then other examples have been found. A further challenge is to arrange the body-diagonal (the internal distance from a corner to the one opposite) also to be a whole number. Such a brick would be called perfect. Unfortunately, no one has yet found a perfect Euler brick—in fact we don't know if one exists.

RELATED THEORIES
See also
NUMBER THEORY
page 30
PYTHAGORAS
page 100
TRIGONOMETRY
page 102

3-SECOND BIOGRAPHIES
PAUL HALCKE
d. 1731
LEONHARD EULER
1707–1783
CLIFFORD REITER
1957–

30-SECOND TEXT
Richard Elwes

3-SECOND SUM
A brick is a shape built from six rectangles. The Swiss mathematician Leonhard Euler was interested in special bricks with dimensions that are all whole numbers.

3-MINUTE ADDITION
Whether perfect bricks exist or not, there are no "small" examples. Using computers, mathematicians have established that if a perfect Euler brick does exist, one of its sides must be more than 1,000,000,000,000 units long. The closest thing found so far is a perfect parallelepiped, built from two rectangles with four parallelograms (like rectangles but the sides aren't perpendicular). This has all dimensions and diagonals as whole numbers.

Everyone knows what a brick looks like. But has anyone seen a perfect brick? Mathematicians haven't.

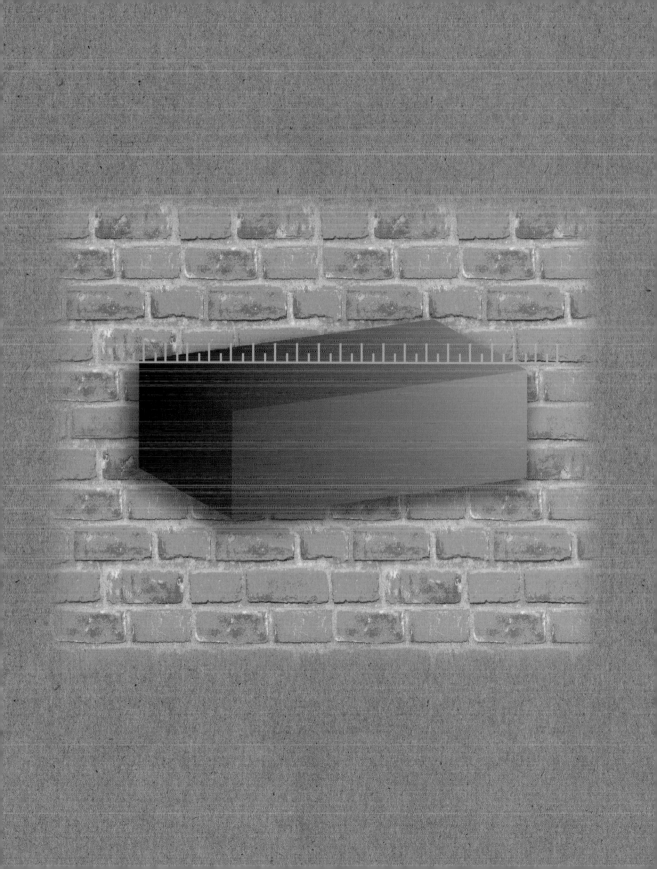

THE MÖBIUS STRIP

the 30-second math

Start with a rectangular strip of paper. Gluing one end to the other produces a cylindrical loop of paper. But if you give the rectangle a half-twist before joining the ends, you end up with something much more exciting: a Möbius strip. This simple paper band's point of interest is that it has only one side and one edge! If you start drawing a line along the centre of the strip, it will cross both the "inside" and the "outside" before reconnecting with itself, since the two sides are actually one and the same. You might wonder what would happen if you cut along that central line. Interestingly, cutting the strip in half does not produce two new loops, but only one. Try it and see! August Möbius' strips have fascinated children and adults since he discovered them in 1858. But for mathematicians, their importance is in the further shapes that can be built from them. If you take two Möbius strips and glue them together along their edges, you produce a single-sided surface known as a Klein bottle. (The only trouble is that it is impossible to create in three-dimensional space, without the surface of the bottle passing through itself.)

RELATED THEORIES
See also
TOPOLOGY
page 116
KNOT THEORY
page 130
POINCARÉ'S CONJECTURE
page 146

3-SECOND BIOGRAPHIES
LEONHARD EULER
1707–1783

AUGUST FERDINAND MÖBIUS
1790–1868

JOHANN BENEDICT LISTING
1802–1882

FELIX KLEIN
1849–1925

30-SECOND TEXT
Richard Elwes

3-SECOND SUM
August Möbius' one-sided loop of paper is a passport to a world of exotic shapes.

3-MINUTE ADDITION
Take a sphere, cut two holes in it, and connect their edges with a cylinder. You have created a torus (a doughnut shape). Take another sphere, cut a single hole, and sew in a Möbius strip along the edge (unfortunately, this is impossible to accomplish in three-dimensional space). It is a fundamental fact of topology that all surfaces can be produced from a sphere through repeating these processes of punching holes and sewing in cylinders and Möbius strips.

A loop with a twist, August Möbius' strip has perplexed and delighted for hundreds of years.

ARCHIMEDES OF SYRACUSE

In popular imagination,

Archimedes is the inventive engineer who ran naked, dripping from his bath through the streets shouting "Eureka!" (I have found it), having discovered a way of determining the volume of an irregular object (by measuring the amount of water it displaces). Like most compelling stories, this is probably not true. But Archimedes did discover what is now called Archimedes' Principle (a law of hydrostatics): the weight of water that a body displaces when immersed in fluid equals the amount of weight it loses to buoyancy. Ancient Greece's best-known practical mathematician is also famous for his eponymous screw pump (based on the lifting properties of the spiral), and his explanation of the principle of the lever. He also invented military weapons, the "claw of Archimedes" (a crane that lifted enemy ships out of the water) and the "heat ray" (a large array of mirrors angled to catch and concentrate the Sun's rays, in an attempt to set fire to a hostile fleet); although it is doubtful that either of these worked.

Although his work was known by Greek scholars, written down in the sixth century CE, and familiar to medieval scholars, until recently modern mathematicians could only extrapolate backward that his inventions were based on sound mathematical theory. It was not until 1906, when the Archimedes Palimpsest manuscript was discovered, that the detail of his theoretical work was brought to light. Some deciphering was achieved in the 1910s, but modern imaging techniques have finally revealed all that is known of Archimedes' methods, showing how close he came to determining the value of π, his method for working out the area under a parabola, the invention of the myriad, and the proof with which he was most satisfied, that a sphere has two-thirds of the volume and surface area of a cylinder of the same height and diameter (including its bases). A sculpted sphere and cylinder appeared on Archimedes' tomb (now lost), which had lain neglected until it was discovered and cleaned up by the orator Cicero in 75 BCE, long after his death at the hands of an overzealous Roman soldier during the Siege of Syracuse.

FRACTALS

the 30-second math

In the late 19th and early 20th centuries, mathematicians devised a variety of constructs that were difficult to understand using the mathematics of the time. The Cantor set is an infinite set of points obtained by starting with a line segment, removing the middle third, removing the middle thirds of the two remaining bits, removing the middle thirds of the four remaining bits, and so on. This process of repeating the same step or series of steps is called iteration, and it lies at the heart of fractals. Early examples include curves such as the Koch and Peano curves, and the Sierpinski triangle, which is related to Pascal's triangle. In the Koch curve (related to the Koch snowflake), each straight-line segment is replaced with four one-third scale segments at each iteration, so that the length of the curve increases with each iteration. Such objects are said to have fractional dimension, for example between that of a regular line and the plane. Applying iteration to simple functions like $x^2 + c$, where x and c are complex numbers (having both real and imaginary parts), and graphing the results in the complex plane yields complicated, beautiful objects known as Julia sets. Benoît Mandelbrot used computers to visualize these sets, as well as the related Mandelbrot set, and developed fractals as a distinct branch of geometry in mathematics.

3-SECOND BIOGRAPHIES
GEORG CANTOR
1845–1918

HELGE VON KOCH
1870–1924

WACLAW SIERPINSKI
1882–1969

GASTON JULIA
1893–1978

BENOÎT MANDELBROT
1924–2010

30-SECOND TEXT
Robert Fathauer

The first four steps in the iterative construction of the classical fractal known as the Koch curve.

ORIGAMI GEOMETRY
the 30-second math

Origami, the centuries-old Japanese art of paper-folding, is inherently geometric. In recent decades, numerous advances have been made involving the mathematics of origami. Huzita, Justin, and Hatori formulated a set of axioms for origami, similar to the way in which axioms have been formulated for geometry. In addition, mathematical theorems addressing theoretical questions about origami have been proven in recent years. Algorithms that aid in finding optimal solutions for the folding of complex figures have been developed by Lang and others, along with computer programs that utilize them. Using these, crease patterns can be produced that indicate the mountain and valley folds needed to create a desired form. While origami traditionally has focused on creating representational forms such as animals and flowers, geometric forms are the primary goal in some modern origami techniques. In origami tessellations a grid of creases is used as the starting point in the creation of geometric forms that often involve repetition. Shuzo Fujimoto is largely credited with starting this branch of origami. In modular origami, multiple geometric modules, each made from a single sheet of paper, are combined to form more complex models.

RELATED THEORIES
See also
ALGORITHMS
page 84
EUCLID'S ELEMENTS
page 94
PLATONIC SOLIDS
page 114

3-SECOND BIOGRAPHIES
SHUZO FUJIMOTO
1922–

HUMIAKI HUZITA
1924–2005

ROBERT LANG
1961–

30-SECOND TEXT
Robert Fathauer

An origami tessellation in which a single sheet of paper has been folded into a repeating pattern of squares.

RUBIK'S CUBE
the 30-second math

The Rubik's Cube was invented

by Ernö Rubik in 1974 and sold in his native Hungary from 1977. In 1980, Ideal Toy Company began selling it worldwide, and today more than 300 million have been sold. A pivot mechanism allows each of the six faces of the Cube to be rotated independently. There are over 43 quintillion (10^{18}) possible arrangements (permutations) of the 26 pieces. Solving the Cube is made easier by memorizing algorithms for accomplishing a desired result, such as cycling three corners without effecting other changes. A move notation developed by David Singmaster allows algorithms to be written down. Singmaster also developed one of the most popular general solutions for the Cube. To mathematicians, the Cube is nothing more than a physical manifestation of an algebraic group. Analysis of the Cube from this perspective shows that it can be solved from any starting position in no more than 20 moves. Only in 2010 was a mathematical proof of this result obtained. The current (mid-2011) world record for solving the Cube is held by Feliks Zendegs at under seven seconds. Variations on "speedcubing" include blindfolded solving, solving the Cube using a single hand, and even with one's feet.

3-SECOND BIOGRAPHIES
DAVID SINGMASTER
1939–
ERNÖ RUBIK
1944–

30-SECOND TEXT
Robert Fathauer

3-SECOND SUM
The Rubik's Cube® is a mechanical permutation puzzle solved by arranging the pieces so that each face of a 3 x 3 cube is a uniform colour.

3-MINUTE ADDITION
In addition to the original 3×3 Rubik's Cube, 2×2, 4×4, 5×5, 6×6, and 7×7 Cubes have also been produced. The number of permutations for the 7×7 Cube is over 10^{160} (1 followed by 160 zeroes!). Other cuboid versions include the $2 \times 2 \times 3$, $3 \times 3 \times 2$, and $3 \times 3 \times 4$. Versions based on the other four Platonic solids, the tetrahedron, octahedron, dodecahedron, and icosahedron have been made, as well. Other polyhedral versions include the rhombicuboctahedron, truncated tetrahedron, truncated octahedron, and stellated cuboctahedron.

In a Rubik's Cube a series of twists is carried out in order to rearrange a scrambled Cube so that each face is a single colour—the number of possible permutations is a mindboggling 43 quintillion!

KNOT THEORY

the 30-second math

As every sailor knows, there are many varieties of knot. All differ based on the number of times the string crosses over and loops around itself. In knot theory, the central question is whether two knots that look different are in fact different. Two knotted loops are judged to be the same if one can be pulled and stretched into the shape of the other, without cutting or gluing the string. The simplest knot of all is called the unknot: a plain unknotted loop. But even this demonstrates a fundamental difficulty: it is easy to make the unknot appear thoroughly tangled and knotted (as anyone who has gone fishing can tell you). A breakthrough came in 1984 with the discovery of the Jones polynomial, which assigns an algebraic expression to each knot. Each knot has one, and if two knots have different Jones polynomials they cannot be the same. This works well for distinguishing a knot from its mirror image, for example, which was previously a difficult problem. However, there is still no known technique that can tell whether any two knots are the same (some knots known to be different have the same Jones polynomials), or even whether any given knot is knotted at all!

RELATED THEORIES
See also
TOPOLOGY
page 116

3-SECOND BIOGRAPHIES
WILLIAM THOMSON
(LORD KELVIN)
1824–1907

JAMES WADDELL ALEXANDER
1888–1971

JOHN CONWAY
1937–

LOUIS KAUFFMAN
1945–

VAUGHAN JONES
1952–

30-SECOND TEXT
Richard Elwes

3-SECOND SUM
Cut a loop of string, tie some knots in it, and then recombine the ends. How can we tell whether two such knotted loops are really the same? This puzzle has perplexed scientists for over a century.

3-MINUTE ADDITION
The mathematics of knot theory is very important in the wider world of science. For instance, DNA strands in our cells are constantly being knotted and unknotted by an army of enzymes. If the DNA becomes too knotted, the cells usually die. Biochemists who want to understand what the enzymes are doing must analyze the resulting knots mathematically.

Knots come in many forms. But it's tough to tell whether two tangles are really the same.

PROOFS & THEOREMS

algebraic number theory The branch of mathematics that deals primarily with the properties and relationships of algebraic numbers (any number that is a root of a non-zero polynomial that has integer coefficients).

axiom A proposition or statement that is self-evidently true or has been accepted as true without proof.

complex number Any number that comprises both real and imaginary number components, such as $a + bi$, in which a and b represent any real number and i represents $\sqrt{-1}$.

decimal number Any number on the counting line that features a decimal point, for example, 10.256

hypersphere A three-dimensional version of a two-dimensional sphere (surface of a globe). It is a compact manifold without boundary or holes. The hypersphere can be visualized only in four or more dimensions. See also *manifold*.

Klein bottle An object with an enclosed surface that has only one side and no edges. A Klein bottle cannot be visualized in three dimensions without self-intersections. It was named after the German mathematician Felix Klein, who first described the surface in 1882.

linear equation Any equation that when plotted on a graph results in a straight line, hence the word linear. Linear equations are made up of terms that are either constants or products of a constant and a variable.

manifold A manifold is a shape in which each region looks like ordinary Euclidean (or real) space. Manifolds exist in every dimension. A curve (for example, a circle) is a one-dimensional manifold, since every small region resembles a one-dimensional line. A two-dimensional manifold is a surface (for example, a sphere) where every patch appears as a piece of two-dimensional plane. A hypersphere is an example of a three-dimensional manifold, since every small region resembles ordinary three-dimensional space. See also *hypersphere*.

Möbius strip A surface that has one continuous side and one edge. It can be made by twisting a rectangular piece of paper and joining the two ends together.

natural number Also known as a whole or counting number, a natural number is any positive integer on a number line or continuum. Opinion varies, however, on whether 0 is a natural number.

nontrivial solution Any solution to a linear equation in which not all of the variables of the equation simultaneously count as zero. A solution arrived at in which all of the variables count as zero is said to be trivial.

prime number Any positive integer that is divisible only by 1 and itself.

proof theory The branch of mathematical logic that describes proofs as mathematical entities in their own right. Proof theory plays a fundamental role in the philosophy of mathematics.

Pythagorean triple Any set of three positive integers (a, b, and c) that follows the rule $a^2 + b^2 = c^2$. The smallest and best-known Pythagorean triple is 3, 4, and 5, since $3^2 + 4^2 = 5^2$.

real number Any number that expresses a quantity along a number line. Real numbers include all the rational numbers (numbers expressible as a ratio or fraction) and the irrational numbers (those numbers that cannot be written as a vulgar fraction, such as $\sqrt{2}$).

theorem A non-self-evident mathematical truth, the truth of which can be established by a combination of previously accepted facts and/or axioms.

torus In geometry, a doughnut-shaped figure.

whole number See *natural number*.

FERMAT'S LAST THEOREM

the 30-second math

RELATED THEORIES
See also
NUMBER THEORY
page 30
EUCLID'S ELEMENTS
page 94

3-SECOND BIOGRAPHIES
PIERRE DE FERMAT
1601–1665

SOPHIE GERMAIN
1776–1831

CARL FRIEDRICH GAUSS
1777–1855

ANDREW WILES
1953–

30-SECOND TEXT
David Perry

3-SECOND SUM
There are no (nontrivial) whole number solutions to the equation $x^n + y^n = z^n$ if $n > 2$. It took more than three centuries for mathe-maticians to prove this simple statement to be true.

3-MINUTE ADDITION
Fermat's assertion has no obvious practical benefit. However, the elusiveness of a proof fired the imaginations of generations of mathematicians. It is easy to argue that the entire field of mathematics called "algebraic number theory" was brought into existence to tackle this single question, and this field has yielded applications of great importance. Wiles' work stood on the shoulders of giants, and his original announcement made the front page of the *New York Times*.

A 17th-century French lawyer and amateur mathematician Pierre de Fermat was working his way through a copy of Diophantus' *Arithmetica*, when he came to a section concerning Pythagorean triples (whole number squares that add up to a square, such as $3^2 + 4^2 = 5^2$). A formula for generating all such triples occurs in Euclid's *Elements*. Fermat claimed that no such triples would be found if instead of squares one used cubes, or fourth powers, and so on. He wrote in his copy of *Arithmetica* that he had a marvellous proof of the claim, but the margin of the book could not contain it. Hundreds of mathematicians spent thousands of hours trying to discover this proof, but at best were only able to show that the equation had no solutions for specific exponents. Fermat himself published a proof for the case $n = 4$ later in his life. Heavyweights like Euler and Gauss also proved special cases. The first sophisticated attempt to resolve the general case for all n was made by Sophie Germain in the early 19th century. Fermat's Last Theorem was really only a conjecture until 1994, when it was finally proved definitively by the British mathematician Andrew Wiles.

Fermat's marginal note was discovered only after his death. Andrew Wiles' first paper on the proof of Fermat's Theorem takes up 108 pages —the margins are empty.

interuallum numerorum 2. minor autem
1 N. atque ideo maior 1 N. + 2. Oportet
itaque 4 N. + 4. triplos esse ad 3. & ad-
hoc superaddere 10. Ter igitur 2.adsci-
tis vnitatibus 10. aequatur 4 N. + 4. &
sit 1 N. 3. Erit ergo minor 3. maior 5. &
satisfaciunt quaestioni.

IN QVAESTIONEM VII.

CONDITIONIS appositae eadem ratio est quae & appositae praecedenti quaestioni, nisi enim caliud requirit quàm vt quadratorum interualli numerorum sit minimo interuallo quadratorum, & Canores vnitem hic etiam locum habebunt, vt manifestum est.

QVAESTIO VIII.

PROPOSITVM quadratum diuidere in duos quadratos. Imperatum sit vt 16. diuidatur in duos quadratos. Ponatur primus 1 Q. Oportet igitur 16 − 1 Q aequales esse quadrato. Fingo quadratum à numeris quotquot libuerit, cum defectu tot vnitatum quod continet latus ipsius 16. esto à 2 N. − 4. ipse igitur quadratus erit 4 Q. + 16. − 16 N. haec aequabuntur vnitatibus 16 − 1 Q. Communis adiiciatur vtrimque defectus, & à similibus auferantur similia, fient 5 Q. aequales 16 N. & sit 1 N. ⅕ Erit igitur alter quadratorum ¹⁶⁄₂₅ alter verò ¹⁴⁴⁄₂₅ & vriusque summa.

OBSERVATIO DOMINI PETRI DE FERMAT.

CVbum autem in duos cubos, aut quadratoquadratum in duos quadratoquadratos & generaliter nullam in infinitum vltra quadratum potestatem in duos eiusdem nominis fas est diuidere cuius rei demonstrationem mirabilem sane detexi. Hanc marginis exiguitas non caperet.

QVAESTIO IX.

RVRSVS oportet diuidere in duos quadratos. Ponatur rursus primi latus 1 N. alterius verò quotcumque numerorum cum defectu tot vnitatum, quot constat latus diuidendi. Esto itaque 2 N. − 4. erunt quadrati, hic quidem 1 Q. ille verò 4 Q. + 16. − 16 N. Caeterum volo vtrumque simul aequari vnitatibus 16. Igitur 5 Q. + 16. − 16 N. aequatur vnitatibus 16. & sit 1 N. ⅕ erit

August 17, 1601
Born Beaumont de
Lomagne, Tarn et
Garonne, France

1620s
Studied in Bordeaux

1631
Degree in Civil Law from
University of Orleans

1636
Appointed Royal
Librarian, Paris

1636
Manuscript of
*Introduction to Plane
and Solid Loci* circulated,
predating Descartes'
La Géométrie

1654
Corresponded with Pascal
on probability theory

1656
Corresponded with
Huygens

1659
*Account of Discoveries in
the Science of Numbers*
sent to Huygens and
Carcavi

January 12, 1665
Died at Castres

1670
Edition of Diophantus'
Arithmetica published
by Samuel Fermat,
with notes by
Pierre de Fermat

1679
*Introduction to Plane and
Solid Loci* published
posthumously in *Varia
Opera Mathematica*

1994
Fermat's Last Theorem
proved by Andrew Wiles

PIERRE DE FERMAT

Thanks to the mystery that

for centuries surrounded his eponymous theorem, Fermat is among the best-known mathematicians to non-mathematicians. Despite making original and important contributions in the fields of geometry, probability, physics, and calculus, and now hailed as the founder of modern number theory, Fermat fiercely guarded his amateur status all his life. He communicated all his ideas and discoveries in correspondence and manuscript form, and eschewed publication in his lifetime, possibly because he did not want the bother of getting his notes and theories up to publication standard. Like his mentor figure, François Viète (1540–1603), he was by day a lawyer, a councillor in the legislature at Toulouse. Keeping out of the academic world ensured that he did not need to demonstrate rigorously his proofs or suffer the indignity of peer-review—indeed, some colleagues muttered darkly that he would not produce his proofs because there weren't any, and that he consistently challenged them with problems too difficult to solve. Fermat riposted by proving that some problems had no solutions.

He was highly regarded by the lions of the day such as Beaugard, Cavanci, and, when he was living and working in Paris for a time, Mersenne. Newton publicly acknowledged that he would not have got to differential calculus without Fermat's pioneering work on curves and tangents, and his advancement of the concept of adequality. He enjoyed a famous correspondence with Pascal, in which the two wrestled with a gambling problem and came up with the principles of probability theory. Fermat also (inevitably) had a run-in with Descartes (surely the most tetchy of mathematicians) about geometric theory, and pipped the philosopher to the post, putting out his own theory a year before Descartes published his; Fermat was right, but Descartes, a man of the establishment, used his influence and connections to blacken Fermat's name and trivialize his reputation. Controversial, brilliant, and enigmatic to the end, Fermat left the world with what seemed yet another insoluble puzzle: his famous, teasing Last Theorem, scribbled as if an afterthought in the margin of one of his textbooks, and unsolved for more than 300 years after his death.

THE FOUR COLOR MAPPING PROBLEM

the 30-second math

3-SECOND SUM

You need only four colors to color in the countries on a map so that no adjacent countries are the same color; why never a fifth?

3-MINUTE ADDITION

The four color theorem is the first major theorem proved with the assistance of a computer. Appel and Haken found a mathematical argument reducing the matter from all possible maps to a property of several thousand particular maps, which a computer could check. The use of this nascent technology sparked a debate, continuing today, about whether computer-assisted proofs should be accepted as valid mathematical proofs.

You've drawn a world map, and you wish to make your map more aesthetically pleasing by coloring in the countries. You decide that any two countries that share a border cannot share the same color. France, Belgium, Germany, and Luxembourg will all require a different color, since each of these four countries shares a border with the other three. So, you will need at least four different colors. Will you be forced at some point to use a fifth color? The four color theorem asserts that you will not. No matter how large or complicated a map you wish to color, as long as each country is a contiguous region, it is possible to color the countries with only four colors. In spite of its simple statement, the four color theorem is extremely difficult to prove. It was only in 1976, 100 years after the theorem was first stated, that US mathematicians Kenneth Appel and Wolfgang Haken found a proof. While four colors are sufficient to color maps on a sphere or plane, this is not the case for maps on other types of surfaces. Mapmakers colouring a torus require as many as seven colors, while on a Möbius strip, six may be needed.

RELATED THEORY

See also
TOPOLOGY
page 116

3-SECOND BIOGRAPHIES

WOLFGANG HAKEN
1928–

KENNETH APPEL
1932–

30-SECOND TEXT

Jamie Pommersheim

When shading a map, only four colors are required to ensure that no two bordering countries have the same color. It took mathematicians a century to prove why a fifth color isn't needed.

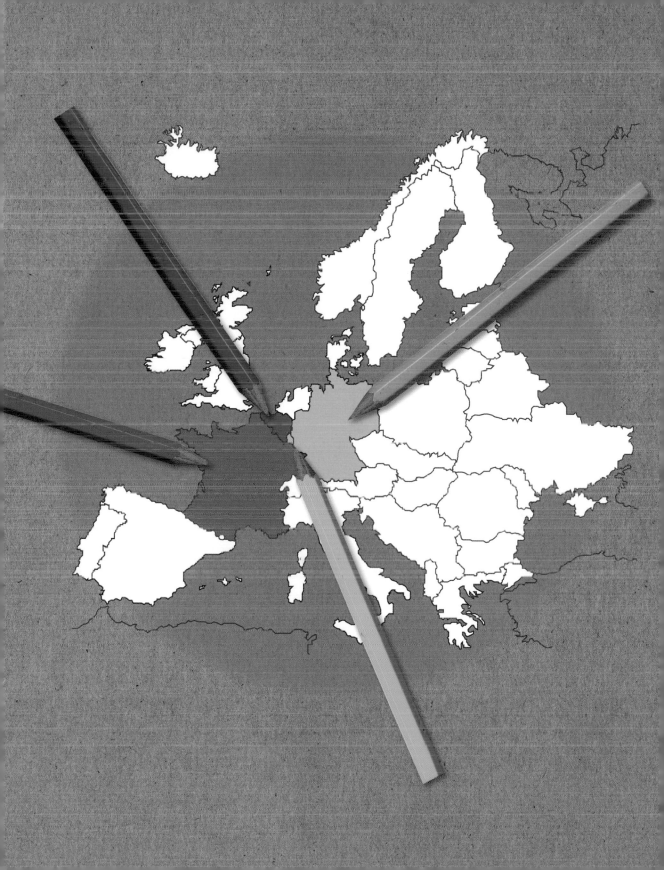

HILBERT'S PROGRAM

the 30-second math

In the early 20th century,
mathematics was in the grip of a "foundational
crisis." While mathematicians were solving
increasingly complex problems, certain basic
questions were left unanswered. Where do
numbers come from? What are their
fundamental laws? Why are some questions
about numbers so extraordinarily difficult?
David Hilbert had a bold idea for addressing
these challenges. He wanted to strip math
down to its bare bones, and treat it as nothing
more than a game. Just as chess is played with
pieces such as pawns and rooks, so the game of
math has symbols as its basic constituents: 0, 1,
$+$, $\times$, $=$, and so on. By reducing math to a
game of symbols, and forgetting what they
"mean," Hilbert sought to discover its
fundamental rules. With this done, he hoped
that an ultimate strategy for winning would
emerge. This would be a single method that
could determine whether any statement about
numbers is true or false. Unfortunately, Hilbert's
program was never realized. Kurt Gödel's
incompleteness theorem showed that a
complete set of rules could never be known.
And later, Alan Turing's work on algorithms
demonstrated that there could never be a single
procedure capable of evaluating the truth of
any mathematical statement.

RELATED THEORIES
See also
ALGORITHMS
page 84
GÖDEL'S INCOMPLETENESS
THEOREM
page 144

3-SECOND BIOGRAPHIES
DAVID HILBERT
1862–1943

WILHELM ACKERMANN
1896–1962

JOHN VON NEUMANN
1903–1957

KURT GÖDEL
1906–1978

ALAN TURING
1912–1954

30-SECOND TEXT
Richard Elwes

3-SECOND SUM
David Hilbert hoped to use
the logic underlying the
structure of arithmetic
to find the ultimate
theory of mathematics.
Unfortunately, his plans
were never to be.

3-MINUTE ADDITION
Although Hilbert's
program failed to meet
his high hopes, his work
had a lasting impact on
mathematics. His
"formalist" approach
of treating numerical
systems as games sparked
new interest in
mathematical logic.
Although a single
computer program or
algorithm can never
solve all mathematical
problems, several special
subclasses of problems can
be resolved this way.
Today's mathematicians
continue to salvage
positive results from
Hilbert's program.

*Like chess,
mathematics is
just a game. But
what are its rules?*

GÖDEL'S INCOMPLETENESS THEOREM

the 30-second math

RELATED THEORIES
See also
INFINITY
page 38
ALGORITHMS
page 84
HILBERT'S PROGRAM
page 142

3-SECOND SUM
Kurt Gödel stunned the world with his revelation that no one will ever be able to write down a complete set of laws of numbers.

3-MINUTE ADDITION
Although Gödel assures us that no complete rulebook for arithmetic can ever be written, a hierarchy of logical systems for arithmetic has subsequently been constructed, where each system plugs many of the gaps of the system below. The subject of "proof theory" compares the logical strengths of these different systems, while "reverse mathematicians" aim to understand where classical mathematical results fit in, asking exactly what underlying assumptions are needed to prove a given theorem.

The centerpiece of mathematics is arithmetic: the system of whole numbers 0, 1, 2, 3, ... together with the well-known ways to combine them: addition, subtraction, multiplication, and division. Mathematicians grappled with this system for thousands of years, and in the late 19th century the focus turned to finding its fundamental laws. What mathematicians sought was a list of the basic rules for arithmetic, from which all higher-level theorems could be logically deduced. Several candidate rulebooks appeared, notably the three-volume work *Principia Mathematica*, by Bertrand Russell and Alfred North Whitehead, which sought to build up the whole of mathematics, starting with a list of fundamental assumptions. However, in 1931 Kurt Gödel proved that all such efforts were doomed. He proved a theorem stating that it is impossible to write down a full list of rules for arithmetic. Any attempt will automatically be "incomplete." There will always be some statement about whole numbers that is missed out: despite being true, it cannot be deduced from the given laws. Of course, you could expand the rulebook to incorporate this statement as a new law, but that would still leave other gaps in the theory. Gödel's theorem guarantees that you can never hope to plug them all.

3-SECOND BIOGRAPHIES
ALFRED TARSKI
1902–1983

JOHN VON NEUMANN
1903–1957

KURT GÖDEL
1906–1978

JOHN BARKLEY ROSSER
1907–1989

GERHARD GENTZEN
1909–1945

30-SECOND TEXT
Richard Elwes

Arithmetic is full of gaps. However many logicians plug, there will always be more.

POINCARÉ'S CONJECTURE

the 30-second math

The surface of a sphere contains no holes. This is obvious. But what does it mean for a surface to have a hole? The mathematical definition is this: If you draw a loop on a sphere, it can be drawn in until it shrinks away to a single point. On a torus (the surface of a doughnut), this does not always work; a loop circling the shape in the right way will get stuck around its hole. For mathematicians, "no holes" means that all loops contract. A double-torus also has holes in it, as does the more exotic Klein bottle. Since the early 19th century, we've known that the sphere is actually the only closed surface without holes, when viewed from the perspective of topology (or "rubber-sheet geometry"). This means every closed surface without holes, such as a cube, can be pulled into the shape of a sphere. Surfaces are two-dimensional shapes. What Poincaré asked was whether the same thing remains true when we step into three dimensions, where surfaces are replaced by shapes called "manifolds." Poincaré believed that the only three-dimensional manifold without holes is the "hypersphere," the bigger brother of the ordinary sphere. This was finally proved in 2003 by Grigori Perelman.

RELATED THEORIES
See also
TOPOLOGY
page 116
THE MÖBIUS STRIP
page 120

3-SECOND BIOGRAPHIES
JULES HENRI POINCARÉ
1854–1912

STEPHEN SMALE
1930–

RICHARD HAMILTON
1943–

MICHAEL FREEDMAN
1951–

GRIGORI PERELMAN
1966–

30-SECOND TEXT
Richard Elwes

3-SECOND SUM
French mathematician Henri Poincaré believed spheres, in all dimensions, to be the only shapes that contain no holes. Over a century later he was finally proved right.

3-MINUTE ADDITION
The Poincaré conjecture can be stated for manifolds in higher dimensions, too. In 1961, Steven Smale and Max Newman proved that in all dimensions from five upward, hyperspheres are indeed the only shapes without holes. Then in 1982, Michael Freedman proved that the same thing is true in four dimensions. So the three-dimensional version, the one that had most interested Poincaré, was in fact the final piece of the jigsaw.

If every loop can shrink away to nothing, then the shape must be a sphere.

THE CONTINUUM HYPOTHESIS

the 30-second math

The list of natural numbers runs on for ever: 1, 2, 3, 4, 5, ... There are also infinitely many real numbers (decimal numbers such as .5 or π or 0.1234567891011121314 ...). These two types of infinity are known as "countable infinity" and the "continuum," respectively. To the dismay of his contemporaries, Georg Cantor proved that these are actually different sizes. In a very real sense, the collection of decimal numbers is a bigger infinity than that of the whole numbers. This was not the end of it: Cantor identified more levels of infinity than these two (infinitely many in fact). But for most ordinary mathematics, these are the two most important types of infinity. Cantor had shown that the continuum is a bigger infinity than the countable level. What he didn't know was whether there were any intermediate levels between them. He believed that there were not, and this conjecture became known as the "continuum hypothesis." It remained open until 1963, when US mathematician Paul Cohen proved the shocking result that the continuum hypothesis is formally undecidable. This means that, given the present set of all mathematical laws, the continuum hypothesis is neither provable nor disprovable.

RELATED THEORIES
See also
INFINITY
page 38
HILBERT'S PROGRAM
page 142
GÖDEL'S INCOMPLETENESS
THEOREM
page 144

3-SECOND BIOGRAPHIES
GEORG CANTOR
1845–1918

KURT GÖDEL
1906–1978

PAUL COHEN
1934–2007

HUGH WOODIN
1955–

30-SECOND TEXT
Richard Elwes

3-SECOND SUM
German mathematician Georg Cantor discovered that infinity comes in many varieties. How these different levels of infinity relate to each other still remains a mystery today.

3-MINUTE ADDITION
Cantor's legacy is one of the few places in which mathematics meets ideology. Cantor's contemporary Leopold Kronecker dismissed the entire subject, saying "God created the integers [whole numbers], all else is the work of man." David Hilbert, on the other hand, declared, "No one shall expel us from the paradise that Cantor has created." These differences of opinion continue today. While some set-theorists search for new laws that would allow the continuum hypothesis finally to be decided, others hold that we can never know.

Infinity comes in different sizes. But how can we know when we've found them all?

RIEMANN'S HYPOTHESIS

the 30-second math

Even today, prime numbers
remain one of mathematicians' main concerns.
The trouble is that they are so unpredictable.
It is very difficult to tell when the next prime
will occur: Sometimes they come thick and
fast (for example, 191, 193, 197, 199), and at
other points there are longer gaps between
them (for example, 773, 787, 797, 809). Yet in
1859, Bernhard Riemann produced a formula
making sense out of this chaos. It was exactly
what mathematicians were seeking. It could tell
the exact number of primes below any limit,
thereby predicting the next prime with complete
accuracy. Although experiments suggested that
it worked perfectly, Riemann wasn't able to
prove that it would always give the right
answer. The formula centered on a mysterious
object, called the "Riemann's Zeta function."
A function is a rule that takes in one number as
input and spits out another as output. In
Riemann's case, this function had both inputs
and outputs being complex numbers (see
Imaginary Numbers, page 18). What Riemann
needed to know was which of the inputs
produced zero. He believed and hypothesized
that all the important zeroes lie on a vertical line
that hits the real axis (a) at $\frac{1}{2}$, dubbed the
"critical line." Yet neither he nor anyone since
has been able to prove for certain that it is true.

30-SECOND TEXT
Richard Elwes

*Do Riemann's zeroes
all lie on the vertical
line at $\frac{1}{2}$? This
question stands
between us and the
mysteries of the
prime numbers.*

Critical line

b

1381 1399 1409 1423 1427 1429 1433 1439

2333 2339 234

a

-2   -1   0   ½   1

3083 3089 3109 3119 3121 3137 3163 3167 3169 3181

3433 3449 3457 3461 3463 3467 3469 3491 3499

APPENDICES

RESOURCES

BOOKS

*50 Mathematical Ideas You
Really Need to Know*
Tony Crilly
(Quercus, 2008)

The Book of Numbers
John H. Conway and Richard K. Guy
(Copernicus, 1998)

The Colossal Book of Mathematics
Martin Gardner
(W. W. Norton & Co., 2004)

Designing and Drawing Tessellations
Robert Fathauer
(Tessellations, 2010)

e: the Story of a Number
Eli Maor
(Princeton University Press, 1998)

Fermat's Enigma
Simon Singh
(Anchor Books, 1998)

*Flatland: A Romance of
Many Dimensions*
Edwin Abbott
(Bluebird Classics, 2013)

Fractal Trees
Robert Fathauer
(Tarquin Publications, 2011)

*Gödel, Escher, Bach: An Eternal
Golden Braid*
Douglas Hofstadter
(Basic Books, 1979)

How To Solve the da Vinci Code
Richard Elwes
(Quercus, 2012)

*Innumeracy: Mathematical Illiteracy
and its Consequences*
John Allen Paulos
(Hill and Wang, 1988)

The Man Who Loved Only Numbers
Paul Hoffman
(Fourth Estate, 1998)

Mathematical Puzzles and Diversions
Martin Gardner
(Penguin, 1991)

Mathematic 1001
Richard Elwes
(Firefly Books, 2010)

*Number Theory: A Lively Introduction
with Proofs, Applications, and Stories*
James Pommersheim, Tim Marks,
and Erica Flapan
(John Wiley & Sons, 2010)

*The Princeton Companion
to Mathematics*
Timothy Gowers (ed)
(Princeton University Press, 2008)

*What Is the Name of This Book?
The Riddle of Dracula and Other
Logical Puzzles*
Raymond Smullyan
(Penguin Books, 1981)

WEBSITES

+Plus Magazine
http://plus.maths.org/content/
An online mathematics journal with
the latest mathematical news and
articles from top mathematicians and
science writers.

Cut the Knot
http://www.cut-the-knot.org/
An encyclopedic collection of maths
resources for all grades. Arithmetic
games, problems, puzzles, and articles.

MacTutor History of Mathematics Archive
http://www-history.mcs.st-and.ac.uk/
Mathematical archive covering the
development of mathematics with
biographies of famous mathematicians.

Math is Fun
http://www.mathsisfun.com/
Math resources for children, teachers,
and parents—with a useful illustrated
dictionary.

The Mathematica Demonstrations Project
http://demonstrations.wolfram.com/
Animations related to a wide range of
math topics.

PlanetMath
http://planetmath.org/
PlanetMath is a virtual community that
aims to help make mathematical
knowledge more accessible.

Wolfram MathWorld
http://mathworld.wolfram.com/
An extensive mathematics resource and
the world's largest collection of
mathematical formulas and graphics.

NOTES ON CONTRIBUTORS

Richard Brown is a member of the faculty and the Director of Undergraduate Studies in the Mathematics Department at Johns Hopkins University in Baltimore, Maryland. His mathematical research involves using dynamical systems to study the topological and geometrical properties of surfaces. Indeed, he studies how the topological transformations of a space affect the geometry of that space. He is also active in studying and enhancing the effectiveness of undergraduate university education in mathematics and how students navigate the difficult transition between secondary school mathematics and university mathematics.

Richard Elwes is a mathematician and teacher. A logician by training, he has published several papers on model theoretic algebra. His books include *Mathemtics 1001* and *How To Solve the da Vinci Code*. He regularly writes on mathematical matters for the *New Scientist* magazine, and enjoys giving talks and masterclasses at schools and in public. He has appeared on the BBC World Service and in the *Guardian*'s Science Weekly podcast. He is currently working as a Teaching Fellow at the University of Leeds, England, where he lives with his wife.

Robert Fathauer is a puzzle designer, artist, and author. He is the owner of Tessellations, a company that specializes in products that combine mathematics and art. He has written articles on Escher-like tessellations, fractal tilings, and fractal knots, and his books include *Designing and Drawing Tessellations* and *Fractal Trees*. He has also organized numerous group exhibitions of mathematical art, both in the US and Europe. He received a BS in Mathematics and Physics from the University of Denver and a PhD in Electrical Engineering from Cornell University. For several years he was a researcher and group leader at the Jet Propulsion Laboratory.

John Haigh is Emeritus Reader in Mathematics at the University of Sussex, England. His main research interests have been in the applications of probability, especially in biology and gambling. As well as teaching in universities, he has delivered popular lectures in series organized by the Royal Statistical Society and the London Mathematical Society. His books include *Taking Chances*, an account of probability for the layman, and (jointly with Rob Eastaway) *The Hidden Mathematics of Sport*, which shows diverse ways in which mathematical thinking can enhance success and enjoyment of sport.

David Perry holds degrees in Mathematics from the University of Wisconsin at Madison and the University of Illinois at Urbana-Champaign. He taught for two years at Ripon College in Wisconsin before becoming a software developer in the private sector. He has also taught for the Johns Hopkins' Center for Talented Youth program every summer since 1997, teaching classes in Number Theory, Cryptology, and Advanced Cryptology. He wrote numerous exercises for the textbook *Number Theory: A Lively Introduction with Proofs, Applications, and Stories* by James Pommersheim, Tim Marks, and Erica Flapan. In addition, he is working on his first novel, a work of historical fantasy that purports to reveal the true story of David and Goliath.

Jamie Pommersheim is the Katharine Piggott Professor of Mathematics at Reed College, Portland, Oregon. He has published research papers in a wide variety of areas, including algebraic geometry, number theory, topology, and quantum computation. He has enjoyed teaching number theory to students at many levels: college math and math education students, talented high school students, and advanced graduate students. He is the co-author of *Number Theory: A Lively Introduction with Proofs, Applications, and Stories*.

INDEX

ACKNOWLEDGMENTS

PICTURE CREDITS
The publisher would like to thank the following
individuals and organizations for their kind
permission to reproduce the images in this book.
Every effort has been made to acknowledge the
pictures; however, we apologize if there are any
unintentional omissions.

129: Rubik's Cube® used by permission of Seven
Towns Ltd. www.rubiks.com

131 Permission to use knot imagery courtesy of Dale
Rolfsen, Rob Scharein, and Dror Bar-Natan.

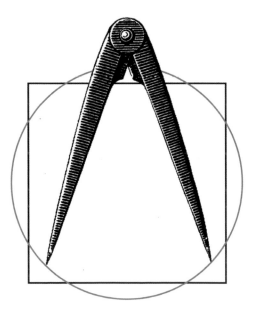